TRANSPORT PROJECTS, PROGRAMMES AND POLICIES

First published 2003 by Ashgate Publishing

Reissued 2018 by Routledge
2 Park Square, Milton Park, Abingdon, Oxon OX14 4RN
711 Third Avenue, New York, NY 10017, USA

Routledge is an imprint of the Taylor & Francis Group, an informa business

A Library of Congress record exists under LC control number: 2002043974

ISBN 13: 978-1-138-70827-3 (hbk)
ISBN 13: 978-1-138-70824-2 (pbk)
ISBN 13: 978-1-315-19854-5 (ebk)

Contents

Notes on Contributors *vii*
Foreword *xiii*
Acknowledgements *xix*

PART I: THE ROLE OF EVALUATION

1 Transport Appraisal in a Policy Context
Peter Mackie and John Nellthorp 3

2 Strategic Transport Planning Evaluation – The Scandinavian Experience
Henning Lauridsen 17

3 Old Myths and New Realities of Transport Infrastructure Assessment: Implications for EU Interventions in Central Europe
Deike Peters 43

4 Norwegian Urban Road Tolling – What Role for Evaluation?
Odd I. Larsen 73

PART II: TECHNICAL ASPECTS OF EVALUATION

5 Spatial Economic Impacts of Transport Infrastructure Investments
Jan Oosterhaven and Thijs Knaap 87

6 The Economic Development Effects of Transport Investments
David Banister and Yossi Berechman 107

7 European versus National-Level Evaluation: The Case of the PBKAL High-Speed Rail Project
Rana Roy 125

8 Welfare Basis of Evaluation
Marco Ponti 139

9 Conceptual Foundations of Cost-benefit Analysis: A Minimalist Account
Robert Sugden 151

PART III: EVALUATION IN THE POLICY PROCESS

10 Impact Assessment of Strategic Road Management and Development Plan of Finnish Road Administration
Eeva Linkama, Mervi Karhula, Seppo Lampinen and Anna Saarlo 173

11 Major Infrastructure Transport Projects Decision-Making Process: Interactions between Outputs and Outcomes as a Contemporary Public Action Issue
Marianne Ollivier-Trigalo 197

12 Involving Stakeholders in the Evaluation of Transport Pricing
José M. Viegas and Rosário Macário 213

13 Accessibility Analysis Concepts and their Application to Transport Policy, Programme and Project Evaluation
Derek Halden 227

14 Strategic Environmental Assessment and its Relationship to Transportation Projects
Paul Tomlinson and Chris Fry 243

Index *265*

Notes on Contributors

David Banister is Professor of Transport Planning at University College London and has recently been a Visiting Fellow at the Warren Centre in Sydney working on a sustainable transport strategy for that city. He has written and edited 16 books on transport planning, the environment and development, including one with Yossi Berechman on *Transport Investment and Economic Development*. Their contribution is based on material from this book.

Yossi Berechman is Professor and Chairman of the Public Policy Department at Tel Aviv University, Israel. His major research interests cover transport economics, transport and land use systems planning, and policy analysis. He has published extensively in these fields and has been a faculty member and visiting scholar in a number of American and European universities and research institutes. He is at present a Senior Research Associate at the University Transportation Center in New York.

Chris Fry, Group Manager – Environmental Assessment, TRL Limited. Chris has expertise in developing and applying appraisal and management tools for a range of decision-making contexts. In particular, he is involved in research and consultancy work focusing on improving environmental impact assessment, strategic environmental assessment and sustainability appraisal methods, and furthering the linkages between them. Chris joined TRL in 2000 from the Environment Agency of England and Wales where he was involved in the environmental assessment of water management projects, and has more recently been working in the transport and construction sectors. Chris is a chartered member of the Institution of Water and Environmental Management.

Derek Halden is a transport planning consultant with 20 years experience of transport planning and project development. Working within research, consultancy, central government and local government. He has prepared national guidance in Scotland on various transport planning issues covering transport appraisal and project planning. He has worked on the planning, appraisal, fund assembly and construction of a wide range of transport projects including major road projects, regional rail development, and local bus schemes. In 1996 he founded the specialist transport planning consultancy DHC and the firm now has a large and growing workload. He has been a member of various advisory committees and national groups on Scottish transport policy and planning issues.

Mervi Karhula (M.sc., Physics) is Planning Manager at Finnish National Road Administration (Finnra). She has worked as Environmental Manager and has recently involved in impact assessment of transport planning.

Thijs Knaap graduated *cum laude* from Groningen University in 1996, specializing in econometrics. He spent the next four years at the university's Department of International Economics as a Ph.D. student. During this period, he published on small-sample properties of dynamic panel estimators as well as central bank behaviour, among others, and went to Brown University as a visiting scholar. His thesis, which is due for publication next year, concerns models of economic geography, the estimation of their parameters and their use as a policy evaluation instrument. His work on the effects of a high-speed rail connection to the northern Netherlands is part of this research. Thijs Knaap is currently at Ocfeb, Research Centre for Economic Policy at Erasmus University in Rotterdam where he works on an applied general equilibrium OLG model of the Netherlands.

Odd I. Larsen is an economist from the University of Oslo. He joined the Institute of Transport Economics in Norway in 1980 with prior work experience from city planning in Oslo. At the Institute of Transport Economics his main areas of work included transport models, road pricing (viz studies of the cordon toll schemes in Bergen and Oslo), pricing and supply of public transport services and cost-benefit analysis. From 1993 he was senior research officer and responsible for development and use of passenger transport models at the institute. In 2001 he became professor in transport economics at Molde University College and senior research officer at Molde Research, but is still affiliated to the Institute of Transport Economics.

Henning Lauridsen is Chief Transport Officer at the Institute of Transport Economics (TOI) in Oslo focusing on international projects. He has more than 30 years experience from transport studies at national, regional and international levels on policy development, strategic planning, appraisal and evaluation. He has been head of various research departments at TOI.

Mr. Lauridsen has extensive experience as project manager and team leader of multi-professional policy, planning and review teams. He has professional experience from most transport sub-sectors.

Mr Lauridsen has worked with transport policy and planning in Europe the last few years and has participated in a numbers of *ex post* evaluation studies in Scandinavia and an evaluation of railway projects in Greece supported by the EU Cohesion Fund. He has worked on institutional reform in the transport sector and strategic transport planning. He has in this respect participated in the EU research task force TRANS-TALK. Parallel to this he has been team leader for a pilot project on transport, welfare and economic development in South-Eastern Europe and for several transport sector evaluations in Eastern and Southern Africa.

Eeva Linkama is Planning Director in Finnish Road Administration. She has experience in developing planning activities in FinnRA as part of the national transport system planning in close co-operation with the Ministry of Transport and Communications. She has also worked in Swedish Road Administration and has actively contributed to The Nordic Road Association. Nominated by the EU Commission she is a member of the External Advisory Group (EAG) for the reseach area Sustainable Mobility and Intermodality in connection with the EU R&D activities.

Rosário Macário, graduated in Management and Organisation in 1987, and received her Masters Degree in Transportation in 1994. From 1977 to 1995 she was Official Flight Operations Officer in a foreign air company that allowed a large experience on planning and operational co-ordination as well as management of diversified working teams. Since 1995, she has been responsible for the organisation and management of the European projects at TIS, developing research studies and projects on transport policy, with special incidence on the problems of legal and regulatory framework of urban transportation and on quality management in transport systems.

Peter Mackie is Professor of Transport Studies at the Institute for Transport Studies, University of Leeds. He has worked at regional, national and international levels on the economic appraisal of transport projects, including contributions to the EU consortium EUNET. With Leeds colleagues, he has recently been involved in preparing the cost-benefit appraisal guidelines for the transport projects of EU accession countries, under the TINA programme, and he is now contributing to a transport appraisal toolkit for the World Bank. He is a member of the UK Government's Standing Advisory Committee on Trunk Road Assessment (SACTRA), which has produced influential reports on Induced Traffic and on Transport and the Economy.

John Nellthorp is a transport economist with research experience both in academia and in consultancy, and a particular interest and expertise in appraisal. He graduated in Economics from Cambridge University in 1993 and with an MA in Transport Economics at Leeds in 1994. At Halcrow Fox in Edinburgh he contributed to road and rail project appraisals, demand forecasting, and original research into the potential for Intelligent Transport Systems in Scotland. The impact of urban light rapid transit on economic development was investigated under contract to the Glasgow transport authority SPT. The economic impact of road infrastructure in peripheral regions was examined in the context of ERDF-supported links in the Western Isles. At ITS since 1996, he has participated in work on behalf of the European Commission and the UK Highways Agency developing appraisal methodologies using CBA, MCA and hybrid structures. He has been involved in the development of the UK New Approach to Appraisal for multi-modal corridors and in the EU 4th Framework projects EUNET and SITPRO. His wider interests include the economics of competition and regulation, and he has undertaken demand forecasting research for Eurostar on London-Brussels flows.

Jan Oosterhaven, born in Gouda, The Netherlands, in 1945. He currently occupies the chair of Spatial Economics in the Departments of Economics and Geography of the University of Groningen and teaches regional economics and transport economics. He has previously held positions as an associate and an assistant professor at the same University since 1970. As a visiting professor he taught urban economics in 1985–1986 at the University of California, Los Angeles, and as a senior consultant he worked part-time for TNO Inro in 1998–1999. He finished his M.A. in Econometrics at the Erasmus University Rotterdam in 1969 and wrote his Ph.D on 'Interregional Input-Output Analysis and Dutch Regional Policy Problems' in Groningen (published with Gower in 1981).

His research focuses on interregional demo-economic analysis, spatial interaction models, evaluation of transport infrastructure, and regional policy evaluations. Besides about 50 popular articles and about 60 research reports, he has written over 90 scientific articles on a wide range of subjects, such as sectoral, regional and urban (economic) policy, housing, impacts of social security, high speed rail, road infrastructure, tourism and agricultural projects. His contributions have been published in the *Annals of Reg. Sc.*, *Env. & Pl. A*, *Econ. Syst. Res.* (of which he was the Managing Editor for five years), *Int. Reg. Sc. Rev.*, *J. of Reg. Sc.*, *Papers in Reg. Sc.*, *Reg. Studies*, *Reg. Sc. & Urb. Econ.*, *Southern Econ. J.* and *Transportation.*

Presently he is vice-president of the International Input-Output Association, Member of the Organising Committee of the European Regional Science Association and Member of the Advisory council on economic and social affairs of the three northern provinces of the Netherlands.

Marianne Ollivier-Trigalo is researcher at INRETS (National Research Institute on Transports and Safety, France), analysing the decision-making processes in the fields of transport policy with a particular focus on major infrastructure projects and networks planning. Her research problematic, based on public policy analysis and political sociology concepts, pays attention on conflicts and problems of co-ordination between actors, notably in terms of territorial, public participation and cognitive dimensions. Her last research report aims at approaching how the French politico-institutional system dealt with the notion of sustainability applied to the transport services planning.

Deike Peters is presently a faculty member at the Department of City and Regional Planning at the Technical University of Berlin. Until November 2000, she was the Director of Environmental Programmes at the Institute for Transportation and Development Policy (ITDP), an international sustainable transport advocacy organization based in New York City. She holds Masters degrees in Urban Planning and International Affairs from Columbia University and a Ph.D. in Urban Planning and Policy Development from Rutgers University. Her doctoral dissertation investigated European Union transport sector infrastructure planning and policy in the context of Eastern enlargement.

Marco Ponti, Professor of Transport Economics and Planning at the Politecnico University of Milan, has experience in transport researches and studies, economic evaluation of projects and transport plans and policies. He has been a consultant for the World Bank and other public and private international agencies; for OECD; for the European Conference of Transport Ministries (both as a discussant and as a rapporteur); for EEC (auditor for the Drive-DG XIII programme, expert for DG-TREN on high-speed railways standardization, and responsible for several researches on transport policies and on simulation models); for the Italian Finance Ministry (regulation of the airport and highway sectors), the Environment Ministry (high speed rail project) and for the Ministry of Transport for the first and the present Italian Transport Master Plans. He has been senior economic advisor for a leading rolling stock manufacturer (for transport studies), and for the Italian State Railway (on investments and industrial policies); expert for the Ministry of Public Works, Economic Advisor of the Italian Minister of Transport from January 1995 till May 1996. Chairman of TRT Trasporti e Territorio Srl. Member of the Board of the Venice Public transport company (ACTV). Member of the Scientific Board of Federtrasporto–Confindustria (Italian Industry Association). He has been responsible or consultant in feasibility studies in Africa (Mozambique, Cameroon, Egypt, Lybia, Tunisia, Ethiopia, Somalia, Zimbabwe, Sudan), in Latin America (Chile, Uruguay, Venezuela, Colombia) and in Asia (Thailand, Indonesia). He has held seminaries in several American and European universities, and teaches in Master courses both for leading transport companies and for public administrations (max 10 righe). He is author of a large number of scientific papers, and collaborates regularly with the main Italian economic newspaper.

Rana Roy, FCIT, FILT, is an independent consulting economist based in London, having formerly served as a government economist in Australia and the UK and as Chief Economist of the Netherlands-based European Centre for Infrastructure Studies. Since 1994 onwards, Dr. Roy has been closely associated with the development of EU transport policy on infrastructure charging, infrastructure investment and the Trans-European Networks, has served as an expert adviser to several EC high-level groups, and has authored, co-authored and edited a number of major studies in the field. He is currently engaged in leading the ECMT/EC research programme on optimal transport taxation, in partnership with the governments of five EU member-states.

Anna Saarlo and ***Seppo Lampinen*** are co-partners of YY-Optima Consulting, which specializes in environmental management of strategic transport and urban planning. They have recently been involved in preparing a guide for assessing environmental and social impacts of strategic plans and programmes of the Finnish Road Administration (*Finnra*), assessing impacts of the trunk road network development plan of *Finnra*, preparing a study of the impacts of rural public transport on regional development and another study of the impacts of rural road network on regional development. Anna Saarlo, M.Sc. in planning geography, is also a graduate student at the University of Helsinki, preparing a post-graduate thesis on social impact assessment of strategic transport plans. She is a member of the environmental committee of the Finnish Association of Consulting Firms, SKOL. Seppo Lampinen has a master's degree in transport planning from the Helsinki University of Technology. He is currently preparing a master's thesis at the University of Helsinki on self-understanding of social impacts of strategic transport plans in *Finnra*.

Robert Sugden is Leverhulme Research Professor of Economics at the University of East Anglia, Norwich. He previously held positions at the University of York and the University of Newcastle upon Tyne. His research uses a combination of theoretical, experimental and philosophical methods to investigate issues in welfare economics, social choice, choice under uncertainty, the foundations of decision and game theory, the methodology of economics, and the evolution of social conventions. His first book, *The Principles of Practical Cost-Benefit Analysis* (co-authored by Alan Williams and published in 1978), was for many years one of the best-known introductions to cost-benefit analysis. Subsequently, his work focused on the scope and limitations of rational-choice theories in both descriptive and normative economics. His book *The Economics of Rights, Co-operation and Welfare* (published in 1986) was one of the first applications of evolutionary game theory in social science, and challenged the prevailing idea that the normative principles on which people act can be modelled in terms of rational choice. With Graham Loomes, he was among the pioneers of behavioural economics, developing psychologically-based theories of decision-making – most famously, regret theory – and testing these theories experimentally. As part of this work, he has investigated the systematic 'anomalies' that are found in stated-preference surveys, and has tried to find ways of adapting cost-benefit analysis so that it is robust to these effects.

Paul Tomlinson has over 20 years experience in environmental assessment and is head of TRL's Environmental Assessment and Policy team. He was recently a member of the DTI Foresight Panel on Environmental Assessment and is currently advising the European Conference of Ministers of Transport on appraisal practice in the UK.

He has prepared a Guidance Manual on Strategic Environmental Assessment for Multi-Modal Studies and is currently modernising UK guidance on the assessment of highways. He has also investigated data needs for SEA. Paul is a recognised international expert on EIA and SEA as applied to transportation having delivered training courses under both the EBRD and PHARE programme. He has been responsible for the environmental assessment of numerous major highway projects in the UK and overseas. Among the projects including a highway in Jordan, the Channel Tunnel Rail Link–Arup proposals, light rail projects and other infrastructure projects.

José M. Viegas, born in 1952, in Lisbon, is Full Professor of Transportation at the Civil Engineer Department of Instituto Superior Técnico, since 1992, and Co-ordinator of the Masters Program in Transportation of that Institute since 1987. He has developed research and consultancy work in several domains of transport analysis and policy, specially in Urban Mobility and in Pricing, at national and European level. Besides leading multiple national projects supplying Government, publicly owned or private companies, he has been responsible for scientific leadership of European projects as ISOTOPE (legal and regulatory framework of urban public transport), FISCUS (costs and financing schemes on urban mobility), PATS (Pricing Acceptability in Transport Systems), DESIRE (Designs for Interurban Road Pricing Schemes in Europe) and participates regularly on several committees, working groups and international conferences, namely for the ECMT – European Conference of Ministers of Transport and the UITP – Union Internationale des Transports Publics. He is vice-president of the Scientific Committee of the World Conderence on Transport Research Society.

Foreword

The Editors

This book contains the selected proceedings of a workshop held in Brussels in November 2000. The purpose of the workshop was to bring together a range of people with different interests and perspectives on transport evaluation – users of evaluations and providers, practitioners and theorists, government officials, consultants and academics. The central focus of the workshop was on the interface between the needs of the users and the capability of evaluation to satisfy these needs.

The workshop was conducted as part of the activities of the TRANS-TALK thematic network, supported under the Fifth Framework of the European Union Programme 'Competitive and Sustainable Growth'. The editors and authors wish to acknowledge the European Commission DG-TREN for financial support under research contract 1999, TN. 10869 and thank project officer Ms Catharina Sikow for her encouragement and support. The views expressed here are those of the authors, and do not necessarily reflect those of our sponsors.

From the TRANS-TALK discussions in this and an earlier workshop, a picture emerged of transport appraisal as seen from a wider public choice perspective. The following sections describe this picture, as the authors saw it. Of course, it is in the nature of any discussion that there are differing perspectives, and in this case there were several clashes of perspective which seemed to open up the heart of the topic – so we comment on these too.

There was agreement that evaluation in the transport sector has a strong technical basis and a strong institutional basis also – more so than in many other parts of the public sector. This is partly due to the characteristics of transport projects, each with specific features, and each requiring a decision on the merits of the case which is then irreversible. It is also due to the public nature of most road and transport infrastructure projects; public authorities need a methodology both for internal planning and decision-making purposes, and for external public consultation and planning inquiry purposes. In some countries, institutional support extends to the creation of standard manuals, and to delegation of decision making from the Ministry of Finance to the Transport Ministry on condition that projects pass acceptance tests at agreed levels. The effect of this has been to create a small transport appraisal industry of civil servants, consultants and academics whose task is to implement and improve appraisal so as to serve the decision-making process. For a review of the origins of this industry, see Foster (2001).

However, it is sometimes argued (see for example Parsons, 2002) that the relatively technocratic focus of transport appraisal creates a weak link between the professional community of modellers and appraisal professionals, and the decision makers and policy advisers who are their clients. These people need the right information, at the right time, and in digestible form, to support the decision they have to make.

In terms of the right information, there is sometimes a tension between the measures of impact which the appraisal system is capable of providing and the concerns of decision-makers acting on behalf of the community. For example, travel time savings have typically been the principal item of monetised benefit for road schemes in developed countries. Decision-makers may be more interested in the effects on transport reliability, or the local and regional economy, which are of more direct social concern, but are much harder to evaluate. This workshop explored how the perspective of transport appraisal can be broadened to address these concerns.

Furthermore, information is only useful if it is timely. The perfect cost-benefit analysis which arrives on the Minister's desk a week after the decision was announced is of strictly limited value. Therefore an important area to get right is the relationship between the appraisal process and the project cycle. This requires discipline on both sides; decision-makers need to see appraisal as a helpful aid rather than as a threat to their untrammelled discretion, while appraisal people need to learn to understand the project cycle and operate within it. Transparency is clearly desirable here – fuzzy processes within which deadlines remain undefined until it is too late to meet them are inimical to good quality appraisal and place at risk the effective use of public money.

Information should be digestible. This seems self-evident, but can be difficult to achieve. It means that there is a strong role for translators and interpreters – people who are qualified to understand the appraisal reports and are capable of presenting the results in ways which aid the decision process. This interpreting role has been under-acknowledged in the past and communication failures may be one reason why appraisal has not been more influential.

The need for this interpreting role has become even more acute because a number of developments have combined to increase complexity. Whereas a generation ago, the standard appraisal context was at the project level, with emphasis on the direct transport impacts, modern transport appraisal has become significantly more demanding. For example, consistency needs to be demonstrated between programme objectives and project performance. Assessment is required of packages of measures and extends to policies such as pricing and regulation. It often needs to be multi-modal, and to consider a wide range of impacts including assessment of economic regeneration effects and environmental sustainability. All this together tests the mental powers of the user of this information.

Appraisal, therefore, has become more ambitious, but is it more effective? In order to answer that, it is necessary to address the following problem areas and their interaction:

- *the level of analysis* – whether the focus is on the project level, the strategic level or the policy level. Even though it can be argued that evaluation methods developed for the project level are readily transferable to the evaluation of policies and programmes, experience shows that this is not the case. Each level requires a subtly different approach and relies on different data;

- *the scope of analysis* – depending on both pragmatic and theoretical considerations, methodologies focus on one or a multiple set of impact types. In the latter case, how the impacts are combined or weighted differs depending on the underlying welfare assumptions of the methods or approach used;
- *the timing* – here it is necessary to distinguish between analysis undertaken to determine the policy mix at the outset of policy or project implementation; assessment exercises undertaken to aid choice between alternatives within a predetermined policy regime; and analysis undertaken for the purpose of monitoring or evaluation performance ex-post;
- *the context of evaluation and of decision-making* – this may differ according to the level, scope and timing of the exercise, but also with reference to the assumptions made about public policy, institutions and societal relationships. The 'value' framework of both evaluators and policy-makers is an important contextual factor that needs itself to be placed under scrutiny.

So, the key questions addressed in this set of essays seem to us to be the following. First, what is the role of evaluation in decision-taking, and how is that role evolving? Secondly, how can progress best be made at the technical level on issues of concern to decision-makers such as the relationship between transport investment and economic performance? Thirdly, how are the broader social processes of decision taking themselves evolving, and what are the consequences for appraisal? The remainder of this introduction comments briefly on these three questions.

The Role of Evaluation

In order to decide whether there is a problem with transport evaluation, and how best to respond to it, one natural starting point is the process of technical evaluation as applied to transport policies and projects. In the next chapter, Mackie and Nellthorp introduce Simon's rational/analytical model. This defines a sequence of stages in the evaluation process starting with problem definition, objective setting, then searching for means then forecasting, appraisal and eventually 'decision' (choice between means), then implementation and ex-post evaluation. Some comparisons are made between this model and real decision contexts at both project and programme level.

An important change in transport evaluation has been the substantial broadening of the geographical scale and the scope of coverage required. Lauridsen summarises how some countries are developing experience of assessing national Master Plans for transport and feeding the lessons from earlier ones into contemporary practice. He emphasises the importance of political and institutional factors and their role in shaping the evaluation process.

Of course a stronger view is that decisions are predominantly political in nature and that appraisal is required in order to support decisions which have already been arrived at informally on other 'political' criteria. Appraisal is then helpful in enabling technical and tactical choices to be made but can be an embarrassment if it challenges the strategy. There are echoes of this in Peters' account of the process of developing the Trans-European Network plans.

Appraisal has been used most extensively in the context of infrastructure investments, but increasingly it is required in a policy context. Larsen's paper provides an assessment of various tolling options for Oslo; again social and political considerations are seen as highly relevant influences on the outcome.

Technical Aspects of Evaluation

The second question concerns the technical 'gaps' in the current state of the art. These are the focus of much research across Europe. Two particular areas are covered in this book.

The first of these is the relationship between transport appraisal and overall economic system impacts. Politicians are interested in the latter, but conventional transport benefit measures do not really consider the wider economic impacts. Are these wider impacts on the economy significant? Are they additional to the measured transport sector benefits, or do they involved double counting? These issues of economic competitiveness and regeneration, and the spatial distribution of output and employment effects, are seen to be of high political importance. Two chapters by Oosterhaven and Knaap and Banister and Berechman consider these issues. Their discussion touches on the concept of 'network effects' and 'community added value' as well as the additionality issues. A different case, illustrating the potential for undercounting of benefits in international appraisals, is considered in the paper by Roy.

The second area is the welfare and public choice foundations of evaluation. Cost-benefit analysis has been subject to criticism from some quarters for a rather narrow perspective which relies on restrictive assumptions and focuses on some of the effects of projects to the exclusion of others. The chapters by Ponti and Sugden each make the case for a renewed understanding of what cost-benefit analysis can and cannot tell us. Sugden takes the welfare-theoretic approach, and examines the scope to free CBA from some of its limiting underlying assumptions. Ponti focuses on the public choice aspects, the central and unique role of social values, and the potential for CBA to be used to avoid the worst excesses of the partisan, discretionary and nakedly 'political' behaviour of politicians with responsibility for transport. This links to one of the main themes in part three.

Evaluation in the Policy Process

On one view, evaluation is a tool for assessing value for money. As such, it has an auditing role, at arm's length from and consciously independent of the decision-making activity. It should be based on a set of principles and practices which have a quality of independence from transient political considerations. Those who see appraisal in such a way often stress its status as an input to quasi-judicial proceedings such as Planning Inquiries and hence to the desirability of maintaining the independence of the appraisal as an 'objective' assessment of the merits of the case as distinct from that of the scheme promoter as an interested party to the case. This can cause problems where different parts of the same agency are responsible for promoting the scheme or strategy and for appraising it.

On the second view, appraisal should be seen as an intrinsic part of the bargaining or argumentation process by which projects are decided. This is the theme of chapters by Linkama, Trigalo, and Viegas and Macario. The crucial question for these authors is the acceptability of the project from the perspectives of the main stakeholders. This implies a need for a more disaggregated approach in which the impact of projects on residents, travellers, environmental interests and other distinct groups are identified. This information becomes an active element in the process by which social trade-offs are made and decisions reached. In this context, the strengths of formal analytical methods such as cost-benefit analysis may become weaknesses, or at least may need to be complemented by a different style of analysis.

Another aspect of this approach concerns again the alleged narrowness of transport appraisal. The final two papers seek for a broadening which could be useful within the context of the stakeholder approach. Halden calls for a strengthening of the role of accessibility analysis within the evaluation, seeing measures of accessibility change as preferable to conventional transport benefit measures. Tomlinson shows how, at the regional policy level, Strategic Environmental Assessment is making a mark as a standardised framework which can tie project evaluation into a wider environmental context.

In reality, many would argue that evaluation must already involve both elements of social audit and argumentation. If correct, this suggests that we need to pay attention to three aspects of the interface between appraisal and decision-making. First, there is the role of appraisal at different stages in the decision process. This includes the familiar conundrum that appraisal is most likely to influence decisions which are genuinely open, yet some evaluation methods require data which is only available right at the end of the process, by which time most options may be politically foreclosed. How to get serious, credible appraisal earlier in the project cycle is a key question. Secondly, different methods of appraisal are appropriate for different types of decision (project, policies, large, small). There are trade-offs here between consistency (one appraisal method fits all) and the innate desirability of treating a large, risky, irreversible infrastructure project appraisal differently from assessment of a small reversible management measure. Thirdly there are practical constraints including the cost and time taken to undertake appraisal itself. The strongest impression from the decision-takers at the Workshop was of the great pressure of time and resources they were under at the latter stages of the decision process, with consequences for their ability to assimilate the technical material. Clarity and comparability of evaluation results are therefore at a premium. Again, there are tensions between the political need for flexibility in the timing and announcement of decisions and the need to undertake a programme of technical appraisal work in a proper efficient manner. The need for mutual understanding and respect between the technical appraisal and decision-making communities of how each is expected to operate within the project cycle is paramount.

Conclusion

In our view, there is great diversity and yet a deep commonality between the approaches to evaluation put forward within the chapters of this book. We consider that this is emphatically positive. There seems to be scope to improve upon the state of the art, whilst continuing to select appropriate forms of evaluation in different contexts, and whilst keeping decision-makers well-informed about what evaluation processes can capture and what the results do and do not imply.

References

Foster, C.D. (2001), 'Michael Beesley and Cost-Benefit Analysis', *Journal of Transport Economics and Policy*, 35, pp 3-30.

Parsons, W. (2002), 'Analytical Frameworks for Policy and Project Evaluation: Contextualising Welfare Economics, Public Choice and Management Approaches', in Giorgi, L. *et al* (eds), *Policy and Project Evaluation in Transport*, Ashgate, pp.144-208.

Acknowledgements

This book derives from the seminar 'Projects, Programmes, Policies: Evaluation Needs and Capabilities' organised by the Interdisciplinary Centre for Comparative Research in the Social Sciences (ICCR) in Brussels in November, 2000. The seminar was organised with the support of the European Commission (DG Energy and Transport), as one element of the work of the Fifth Framework Programme Thematic Network TRANS-TALK on Policy and Project Evaluation Methodologies in Transport.

The editors would like to thank the European Commission for its support of TRANS-TALK, and particularly Catharina Sikow-Magny of DG-TREN for her continued supervision of our work. We wish to thank also all those who attended the Brussels seminar for their contributions to a lively and on-going debate.

We are indebted to Liana Giorgi and Annuradha Tandon of the ICCR for co-ordinating the series of TRANS-TALK seminars and to Rudolfine Gamboa, also of the ICCR, for her assistance with copy editing and the preparation of the final manuscript for the publishers.

PART I
THE ROLE OF EVALUATION

Chapter 1

Transport Appraisal in a Policy Context

Peter Mackie and John Nellthorp

Introduction

In the first TRANS-TALK seminar Wayne Parsons gave a stimulating paper in which he threw down the gauntlet to the appraisal community (Parsons, 2001). He claimed that, relative to other policy areas, transport appraisal is technocratic in nature, perhaps excessively so. As a result, there is a danger of communication failure; decision-makers do not understand what technical appraisal outputs mean, while technocrats do not appreciate what decision-makers need in order to make decisions. This raises a number of interesting questions. Is this portrait true? What is the appropriate role of technical analysis within the decision-making process? Is there a tension between improving the technical quality of appraisal and improving the relevance and clarity of appraisal information for decision-takers? As we noted when presenting this paper at the second TRANS-TALK seminar, presumably most of the audience believed that technical analysis should be a relevant input to decision taking. So what challenges do we face in trying to deliver appraisal which is technically improved and more useful to decision-takers? Broadly speaking, that was the agenda for the second TRANS-TALK seminar, and it is also the agenda for this book.

This is a timeless agenda. For example, it is very reminiscent of the debate over the role of cost-benefit analysis in public policy making three decades ago, at the time of the third London Airport Inquiry, when economic analysis sought to intrude rudely on more traditional methods. That debate was captured in an entertaining contemporary literature – see Self (1970, 1975) and Williams (1973).

Our view of the transport decision process today is perhaps a little different from that of Parsons. We see it as a world of overlapping influences, see Figure 1.1.

Probably the situation is not the same in all countries and for all types of projects, but we see three broad sets of influences on whether a project goes ahead:

- *What the public thinks*. This influence has become more powerful in the UK over the last decade. For example, in many situations, if local people oppose a new by-pass for their town, it will not happen regardless of the economics. Again, public opposition to road user charging remains a significant barrier to implementation;
- *What the economic appraisal says*. This is the economic appraisal in its broadest sense, that is, including environmental and wider impacts and showing the incidence of effects on specific groups. The results may be given

within a Framework or Multi-Criteria Analysis, which seeks to summarise the key relevant information;

- *The political context.* This includes the extent to which the project is consistent or conflicting with the overall goals of the administration, whether it fits with prevailing philosophy e.g. pro-market/pro-regulation, how strongly the relevant lobby groups want or oppose it, and in a few cases, naked political advantage.

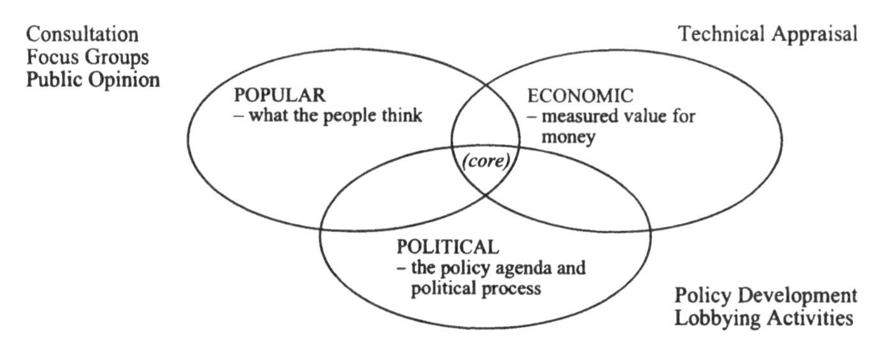

Figure 1.1 Influences on Decision-Making: the Three Spheres of Appraisal

We can think of a lot of the processes within the project cycle as trying to find solutions, or packages, which lie within the core of Figure 1.1 – that is, solutions which are acceptable in social, economic and political terms. For example, a highways authority might design the route for a major project. Then at public consultation, that route might draw the response that it lies too close to a particular town. Then the scheme might be redesigned, at a cost penalty. Another example is urban road pricing, which is currently under consideration for a very few cities in the UK. Urban road pricing can perform well in economic terms (e.g. Halcrow Fox, 1999), but will need to be packaged with other measures in order to gain social and political acceptance (Schade, 2001).

So, the decision taker must consider three questions. Is a project or policy acceptable to the public? Is it economically acceptable? Is it politically acceptable? These are the hurdles which projects and policies have to jump. Usually, projects have to make several circuits of the racecourse. For example, there may be a stage where the initial idea gains acceptance, then a screening process involving outline design and assessment and public consultation. Some projects and options fall at this fence. Then there follows detailed design and cost-benefit analysis. There may be a public inquiry. There may be political lobbying. Finally there is the Ministerial decision, often carefully timed. The most controversial road scheme in the UK of recent times, the Newbury by-pass, was announced by the Minister on his last day in office at Transport.

Crucial to good quality decision-making is an ability to understand, incorporate and balance-off the social, economic and political considerations so as to reach the core of Figure 1.1. This in turn requires appraisal information to be comprehensible, and information overload to be avoided. We think that it is timely to revisit the agenda of Self and Williams because in recent times the appraisal context has changed, and has become more complex in a number of respects:

- the paradigm for appraisal used to be that of the single project for a single mode. Now the emphasis is more on plans at the area or corridor level, and on packages of pricing, regulation and investment measures;
- the Government-led project using public finance is becoming less common, whilst public-private partnerships are in the ascendant, necessitating evaluation from private as well as social viewpoints, and involving many levels of Government (EU, national, regional);
- the balance of power between Governments and the public has shifted. Governments need these days to be able to make their case to an often sceptical public with access to professional expertise;
- the boundaries between technical appraisal and political and professional judgement have shifted. There is a strong desire now to incorporate environmental and economic development impacts within technical appraisal;
- related to all of the above is an overarching wish to relate appraisal more securely to the overall objectives of transport policy. This raises questions of consistency between objectives and evaluation criteria.

In this paper, therefore, we look at conceptual models of appraisal and decision-making. We then examine four documented examples of transport decisions, where the interface between the appraisal and the decision is more transparent than usual, or has been analysed in some depth. Finally, we set out some 'challenges to appraisal people', for the purposes of discussion. These are based on our perception of the weaknesses in current practice and the needs arising from the current policy context.

The Appraisal Process

Two alternative conceptual models of the way appraisal interacts with decision-making are set out in Boxes 1 and 2 (Figure 1.2). The first, due to Simon (1957) stresses analysis as the core activity within appraisal. His methodology is widely taught to engineers, and is clearly highly relevant for single-shot investment decisions involving high sunk cost and significant risks. It is the model many appraisal people instinctively think within. But it is not the only model in the policy analysis literature. The muddling through model of Lindblom (1959) stresses that many decisions involve incremental movement, are reversible, involve bargaining and mutual adjustment and are satisficing rather than optimising decisions. Many pricing and regulatory decisions by public agencies are better described by this model. It is often pointed out that the two models are to a degree complementary, with Simon describing the analysing function, and Lindblom the negotiation function, inherent in decision-making. The interest here is in the nature of the interaction between the two. For a review, see Hogwood and Gunn (1984).

For transport project appraisal, the core of the Simon model can be represented by four activities. A key issue is how these are related to each other, and to what extent they are technical and to what extent political in nature.

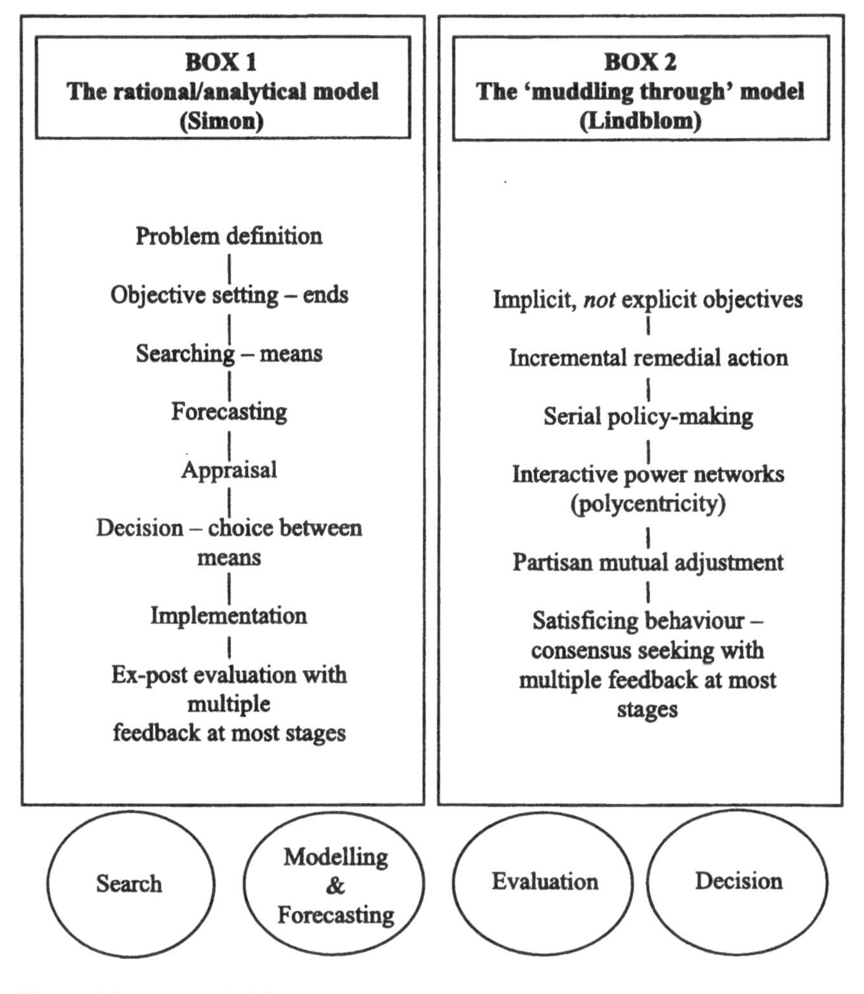

Figure 1.2 Appraisal Process

Search is the domain of ideas generation, policy creation, option development for testing. In many ways it is the most important and yet least understood of the four processes. It is important because unless a project option is developed at the search stage, it cannot be tested and cannot be implemented. It is clearly not a purely technical process. Ask where a project idea comes from, and the answer is as likely to be from the political realm as from the technical one. There is a good book to be written about how projects are created, whether the Channel Tunnel or the Øresund Link, or the Messina Straits Bridge, or the Humber Bridge or the Manchester Metrolink. What options are created for appraisal and which are excluded, what constraints are placed on the appraisal and why, and how the hurdles are specified are political questions.

The *modelling and forecasting* stage is the most technical of the four, but it is not purely technical. In the context of income growth, what is to be assumed about exogenous traffic growth? Is induced traffic to be allowed for? How large is the study area to be? When can the model be considered to be satisfactorily calibrated and validated? The technocrats are likely to be operating within a framework which is set by prior policy and administrative decision. The main role of bodies like SACTRA in the UK, or processes like OEEI in the Netherlands, is to review some part of the framework, and to make recommendations for change.

The *evaluation* stage is similar. It is essentially a technical process, but operating within a set of administrative rules, which make up the evaluation framework. What is the scope of technical appraisal? Should it encompass social exclusion or economic development impacts, for example? Which impacts are assigned monetary values? How are the non-monetised impacts handled? Are values of time and safety willingness-to-pay based values, or are they adjusted to social values by some form of averaging? These are questions which lie at the junction between the technocrats and the politicians, who must decide not only the quality of the evidence and arguments, but also bear in mind the extent to which they wish to preserve a degree of freedom of manoeuvre for themselves.

As argued above, the *decision* stage is where the outcome of technical appraisal is mixed with social and political judgement. It is here that the questions posed previously arise: are the technocrats providing the decision-makers with the information they need, and are the decision-makers specifying clearly what appraisal information they want? Despite occasional moments of depression when policy decisions inexplicable on the basis of appraisal evidence are made (the resurrection of the tram at the expense of the guided bus is a current UK example), our overall perspective is an optimistic one. We believe technical appraisal does make a significant contribution to decision-making. A good test is the counter-factual – in the absence of technical appraisal, would the quality of decisions be different, and if so, better or worse? We tend to believe that the existence of the appraisal system has helped to avert some disasters and on average has improved decision quality. In the next section of the paper, we discuss the relationship between appraisal examples and decision-making.

The Relationship between Appraisal and Decision-Making: Some Examples

What evidence is there to help us understand the relationship between appraisal information and decision-making in practice? How are the political, popular and economic influences balanced? How do the theoretical models in Boxes 1 and 2 relate to real-world processes?

In this section we examine four examples of transport decisions. All were taken when appraisal information was available – we focus on the way in which this information was used, and the other influences which acted on the decision. The examples are:

- the East London River Crossing;
- Vagverket 10–year Investment Programme, Sweden, 1984;
- British Rail Closures, 1963–70;
- the UK Roads Review, 1997/8.

East London River Crossing

The first example is at the project level. The East London River Crossing was a very large bridge project across the River Thames, which was intended to join the road networks north and south of the river at a point near Greenwich. A major Public Inquiry was held in 1985/6. The scheme was proposed by the Department of Transport and opposed by many objectors including the London Borough of Greenwich. The project was marginal in terms of its economics (in a narrow CBA assessment), strongly negative on environment, and argued to be strongly positive on economic regeneration of Docklands – although in the light of the modelling work there were reasons to doubt this.

It was clear during the Public Inquiry that the efficiency and environment issues were not really of central concern to the scheme promoters. There existed a policy context within which the evaluation was subsidiary, namely a political commitment to develop the scheme as a contribution to the wider Docklands strategy. In this case, the technical merit of the scheme played only a limited role in determining the outcome of the Public Inquiry, and the Inspector ruled in favour of the project.

As a footnote, it is worth noting that this scheme was one of those later referred to the European courts for possible breach of Directive 85/337 on environmental impact assessment. By the time a ruling had been obtained, the UK was deep in economic recession, and the scheme was cancelled. A different version of the scheme is now under consideration as the River Thames Bridge.

Vagverket 10–year Investment Programme

In 1984, Vagverket, the Swedish National Road Administration, decided upon a 10–year investment programme, consisting of many distinct infrastructure projects across the national network. The average investment cost of an individual project was approximately 4 million Euro at 1985 prices. The projects were spread across the 24 Swedish 'Counties', with typically 20–50 projects in each county.

At that time, Vagverket followed a formal planning process whereby a Project Analysis (PA) was carried out for each project that a county put forward. The projects were then ranked twice, first by the county, secondly by the Main Office of Vagverket, both of whom had the PA results available. Nilsson (1991) used a econometric model to analyse Vagverket's behaviour in ranking projects. He used the rankings – first the county rankings, then the Main Office rankings – and the PA information as data with which to calibrate the model.

Nilsson tested two hypotheses:

(i) that the observed rankings can be explained by referring to the Project Analysis results;
(ii) that the rank of a project increases (i.e. a better ranking is given) as the economic rate-of-return, predicted by the Project Analysis, increases.

The Project Analysis was basically a cost-benefit analysis, so a key variable in the model was the Net Present Value: Cost ratio. Other explanatory variables included a dummy variable for 'system effects' (to allow for complementarity of schemes) and dummy variables indicating whether the scheme involved a local, national or European-designated road.

Nilsson's results are interesting regarding both the county-level decisions and the national-level decisions. At the county level, he found that hypothesis (ii) was not rejected for five out of seven counties modelled, whilst hypothesis (i) was not rejected for three out of seven. Jointly, (i) and (ii) were not rejected for two counties – i.e. the cost-benefit results were shown by Nilsson's relatively simple model to play a positive role in determining investment priorities in two out of seven counties.

By contrast, at the national level, he found that neither hypothesis could be supported. In other words, it had not been possible to show – using his particular model – that the national-level decisions were related to the appraisal information. In addressing the failure of the model to explain national-level decisions, Nilsson posits a number of alternative explanations based on his knowledge of the actual decision processes involved. The first of these is that the government may be investing in roads in order to alleviate unemployment, although having set up dummy variables for 'supported areas' he finds that the signs on these dummies are sometimes of the wrong sign compared with his expectations. This is tantamount to suggesting that a key policy-relevant impact of these projects has been omitted from the appraisal. Secondly he introduces the notion of historical fairness between the Swedish counties as a factor in allocating funds. If true, this certainly could help to explain how multiple county-level rankings, when merged into one national ranking, become impossible to explain based on NPV/Cost ratios. Thirdly he points to differences between the preferences of the decision-makers involved.

British Rail Closures, 1963–70

Brent (1979) aimed to find – by *ex post* analysis of observed data – the 'weights' placed on various different categories of costs and benefits when decisions were made on railway closures over the period 1963–70. That these decisions were taken in the knowledge of cost-benefit analysis results made it possible to regress a decision variable (0=closure; 1=no closure) on various components of the cost-benefit analysis.

Brent's findings again addressed a narrow range of impacts, and the model variables were set up to capture first and foremost the distributional impact on different groups (rail users vs. road users and high vs. low income groups). The main explanatory variables were in fact rail users' travel time, fare differences between modes, congestion avoided and a variable representing the loss to disadvantaged groups (equivalent in some cases to the size of the financial subsidy which would be required if the service were to remain open). His principal conclusions were that:

- there *was* a difference between the weights placed by decision-makers on impacts on different groups – i.e. distribution *was* a consideration;
- however, the difference between the weight on car users' benefits and rail users' benefits was the opposite of that expected – car users' benefits appeared to be weighted much more heavily in the overall assessment.

It may be possible to learn a lot more about decisions like this – after the fact – by discussing the decision process with the decision-makers themselves. Unfortunately, we do not know why the UK Ministry of Transport placed more weight on car users' benefits than on rail users' benefits, and we can only speculate that the political climate at the time, and in particular the prevailing view of public spending on supported transport services, was not favourable. It would have been interesting to know more.

The UK Roads Review, 1997/8

Finally, we come up to date by looking at the 'Roads Review' carried out on the initiative of the incoming Labour government in 1997. This has since set the framework for appraisal across all modes of transport in the UK.

The Roads Review focused on 68 road schemes – these were selected from a larger 'programme' on the basis that they could all be implemented within the next seven years. Schemes were considered for entry into a Targeted Programme of Improvements (TPI) for which a budget was available. The eventual cost of the TPI was approximatcly two billion Euro.

Each of the 68 schemes was subject to a new form of appraisal, which built on the existing economic appraisal and environmental assessment techniques. The presentation of the appraisal results included a new one page summary, designed specifically to be shown to decision-takers. This Appraisal Summary Table (AST) shows a scheme's performance against each of the Government's five objectives for transport policy: environmental impact; safety; economy; accessibility and integration. An example is given in Annex 1, so the reader can see the mix of qualitative and quantitative information given, and the way the information is summarised. The 68 ASTs were the main source of information available to decision-makers when the Review itself began.

It may be worth mentioning one or two features of the decision process itself. Firstly, the process of decision-making in the Roads Review involved a single team of civil servants with roads policy responsibilities examining all the cases together with the relevant Minister. Reasonable consistency of judgement could therefore be expected. Secondly, as part of a deliberate move to a more open and transparent process of decision-making, the Appraisal Summary Tables were published both in print and on the internet.

As in Nilsson's study in Sweden, Nellthorp and Mackie (2000) used an econometric model to investigate the relationship between the Roads Review decisions and the appraisal information. However, this time the appraisal information was much more wide-ranging, offering the opportunity to capture many more variables within the model.

Based on the modelling work carried out, the following conclusions were drawn:

- there *is* a pattern relating the decisions taken to the information in the ASTs;
- in particular, it was possible to show that the decisions were related to performance on Noise, Landscape, Heritage, Safety, Journey Times, Reliability, Regeneration and Scheme Cost;
- also, the chances of the scheme being accepted were improved where it was clearly demonstrated in the AST that alternative options had been considered and rejected;
- implicit values were inferred for a number of variables (by comparison with Scheme Cost), including notably large values for Reliability, Landscape and Regeneration;
- compared with the standard cost-benefit values for travel time and safety, it appeared that decision-makers had slightly uprated the value of safety and downrated the value of time;
- decision-makers had been presented with monetised and non-monetised information on time savings in the AST – the non-monetised information appeared to be related to the decisions, but it was not possible to find a relationship with the monetised information (although that may have been because the 'true' functional form was not found and tested).

The stated aim of the AST is to make the appraisal process more transparent and to provide a clearer, more systematic and consistent basis on which to found decisions. This research suggests that these efforts may have paid off – there is some evidence from this study that the decisions taken were influenced by a broad range of the appraisal information.

Conclusions

The evidence we have cited is limited, much of it is related to the UK and only some of it is backed up by quantitative analysis. We have major reservations about attempting to generalise it to other decision situations. However, these examples make a start by providing some evidence of the existence of the link between evaluation and policy-making which this book discusses. They also begin to indicate what types of appraisal information are taken into account by real-world decision-makers, and how.

One overall conclusion is that the relationship between Simon's model of rational/analytical planning and the wider political and popular influences on decision-making is a complex one. As political agendas change, priorities change within the transport sector. The East London River Crossing may have been approved at Public Inquiry in 1985/6 when regeneration of Docklands was a top priority, but it was dropped in the nineties when recession hit and priorities lay elsewhere. We do not know what role the appraisal information played in the latter decision, but perhaps it was greater than in the former.

Another conclusion is that in some cases, there is evidence of two-way communication between decision-makers and analysts, with encouraging results. Figure 1.3 crudely summarises this in the UK Roads Review example: decision-makers set out to influence the scope of the appraisal and the way the results were presented; analysts then provided information which the decision-makers appear to have been able to use.

Communication:

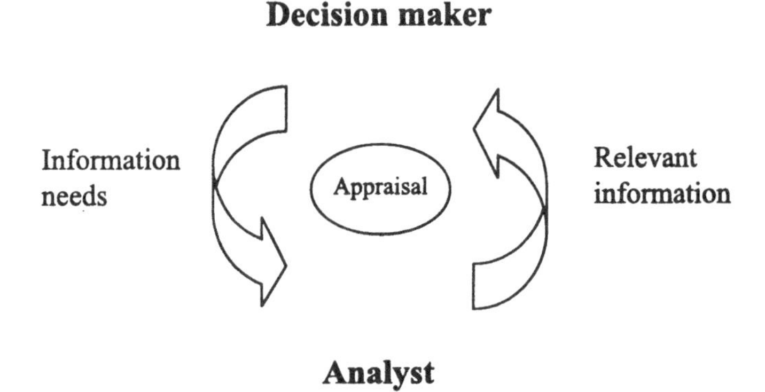

Figure 1.3 Communication between Decision-Makers and Analysts

In the final section, we consider some specific challenges faced by the appraisal community in delivering appraisals which are both technically improved and more relevant to decision-makers.

Five Challenges for Appraisal People

Our overall view is that appraisal is making progress. In the UK context the New Approach to Appraisal and the *Guidance for Multi-Modal Studies* and the *Local Transport Plan Guidance* all emphasise the need for consistent application of appraisal principles within a coherent framework so as to demonstrate value for money in social terms. This places the onus on the appraisal community to respond by improving the relevance and clarity of what it does. There are five challenges.

Valuation

The shape of the cost-benefit analysis used in different member states, in terms of the inputs and outputs which are assigned money values, the basis for these values, and the treatment of non-monetised items is extremely variable. Why do some countries place monetary values on noise and air pollution while others do not? Are values for these items known with an acceptable level of certainty, say with comparable certainty to the value of safety risk, which is monetised in all countries? At a more philosophical level, should the values in cost-benefit analysis be unadjusted willingness-to-pay values? Or should they be adjusted to social values perhaps using a standard or equity value approach, as in the UK? Perhaps the answer to this depends in part on the roles assigned to economic, social and political assessment within the overall appraisal regime.

Double Counting

Under this heading we place all of the issues of overlap between the direct transport impacts of projects and their wider economic impacts. There is little doubt that:

- it is far easier for transport appraisal to model and evaluate the direct transport benefits, but that;
- the wider economic impacts are sometimes of far greater interest to the decision-makers.

Therefore it is essential to keep a cool head in determining how the transport benefits map onto the final economic benefits, and in what circumstances there is evidence that the final benefits are larger than the primary benefits (i.e. there is genuine additionality). This is the area covered by the UK SACTRA (1999) and Dutch OEEI (2000) reports. As with the valuation issues, it is critical to be clear about the assignment of responsibilities between the technical appraisal and the political decision-making process. Is the technical appraisal seen as in principle, a complete assessment, on which the political decision process is then a different perspective? Or is part of the appraisal (say the direct traffic, economic and environmental impacts) the domain of technical appraisal, with the political process taking responsibility for assessing the wider impacts through other means? Our view is that it should be the first, but then the technical appraisal people have a lot of work to do to produce a convincing integrated transport and the economy assessment.

Distributive Impacts

Many of the above remarks apply even more forcefully when we come to consider the distributive impacts of transport projects. Who really gains and who loses? It depends on the transmission system and on market characteristics especially for land and labour, but also for goods. There is a long way to go here. One of the UK Government's objectives is to reduce social exclusion. But in the context of appraisal of road and transport infrastructure there is little concept of how the appraisal should handle this, even at the level of indicators or qualitative assessment.

Multi-stakeholder Evaluation

Thirty years ago, life was simple. Everybody knew what cost-benefit analysis meant. It meant the assessment of costs and benefits to society, whoever the benefits accrued to. Now we have moved to a different form of social organisation, with hierarchical and overlapping Government structures (EU, national, regional, local) and public-private partnerships. This places a lot more strain on the evaluation system, in two ways.

First, each layer of Government needs to assess the quality of the project from its own perspective. What are the benefits and costs viewed from where it stands? Are there any special benefits for which it has a particular responsibility or interest and for which it should be willing to make a contribution? In the particular case of the EU it is important to address the issue of network effects and European Added Values. What are these, and are they really additional to the national CBA, and under what circumstances?

Secondly, many, or perhaps most, projects involve significant private sector interests. Especially for public transport and toll motorway projects, the form of the pricing and franchise arrangements may be interdependent with the overall social evaluation, and it is necessary to have an evaluation tool which can display the impacts on relevant parties and permit analysis of the trade-offs between revenues, user benefit and costs.

Communication

Overlaying all of the above points is the need to foster communication between appraisal people and decision-takers. In that context, there is a tension between greater sophistication in appraisal and risk of communication failure. The appraisal community tends to underrate the difficulties of reaching their natural consumers, the decision-takers. So, as well as finding ways of improving the quality of the technical appraisal, we also need to find ways of communicating its meaning effectively in the decision process.

References

Brent, R.J. (1979), 'Imputing weights behind past Railway Closure Decisions within a Cost-benefit Framework', *Applied Economics*, Vol. 11, pp.157–170.

Central Planning Bureau, Dutch Economic Institute (2000), *Appraisal of Infrastructural Projects: Guide for Cost-Benefit Analysis* (the OEEI report), Dutch Economic Institute.

Department of the Environment, Transport and the Regions (DETR) (1998), *Understanding the New Approach to Appraisal*, London, DETR.

Grant-Muller, S.M., Mackie, P.J., Nellthorp, J. and Pearman A.D. (2001), 'Economic Appraisal of European Transport Projects – The State of the Art Revisited', *Transport Reviews*.

Halcrow Fox (1999), *Road Charging Options for London: A Technical Assessment*, The Stationery Office, London.

Hogwood, B.W. and Gunn, L.A. (1984), *Policy Analysis for the Real World*, University Press, Oxford.

Lindblom, C.E. (1959), 'The Science of Muddling Through', *Public Administration Review*, Vol. 19.

Nellthorp, J. and Mackie, P.J. (2000), 'The UK Roads Review: A Hedonic Model of Decision-making', *Transport Policy/WCTR-S Journal*, Vol. 7, p.2.

Nilsson, J.E. (1991), 'Investment Decisions in a Public Bureaucracy: A Case Study of Swedish Road Planning Practices', *Journal of Transport Economics and Policy*, May, pp.163–175.

OEEI (2000), 'Appraisal of Infrastructural Projects: Guide for Cost-Benefit Analysis', Central Planning Bureau, Netherlands Economic Institute.

Parsons, W. (2001), 'Analytical Frameworks for Policy and Project Evaluation: Contextualising Welfare Economics, Public Choice and Management Approaches', in L. Giorgi *et al.*, *Policy and Project Evaluation in Transport*, Ashgate, Aldershot.

Schade, J. (2001), 'Acceptability of Marginal Cost Pricing in Urban Transport', Paper presented at the NECTAR Conference, Espoo, Helsinki, 16–18 May.

Self, P. (1970). 'Nonsense on Stilts: Cost Benefit Analysis and the Roskill Commission', *Political Quarterly*, Vol. 41, No. 3.

Self, P. (1975), *Econocrats and the Policy Process: The Politics and Philosophy of Cost-Benefit Analysis*, Macmillan, London.

Simon, H.A. (1957), *Models of Man*, John Wiley, London.

Standing Advisory Committee on Trunk Road Assessment (SACTRA) (1999), *Transport and the Economy*, Report to the Department of the Environment, Transport and the Regions, London: The Stationery Office.

Williams, A. (1973), 'CBA: Bastard Science and/or Insidious Poison in the Body Politick?', in J.N. Wolfe (ed.), *Cost-Benefit and Cost Effectiveness Analysis*, pp.30–63.

Annex 1: Appraisal Summary Table (AST)

Version of 22 July 1998

A6 Clapham, Bedford (GOER)		1996 scheme - 5km D2 bypass		Cost £30.9m
PROBLEMS		Poor safety and environment within Clapham (pop 3,200) where A6 carries up to 21,000 vpd (7% HGV). 300 residential properties + 2 schools front on to the road. Peak hour queuing occurs on length between village and northern outskirts of Bedford.		
OTHER OPTIONS		2 Pelican crossings already provided in village. Large scale traffic calming scheme would have unacceptable noise, air pollutn and severance effects. Other options considered include reduced standard single carriageway bypass on proposed line; an eastern bypass and improved rail services to new Bedford North station with park and ride. All have inferior benefits.		
CRITERIA	SUB-CRITERIA	QUALITATIVE IMPACTS	QUANTITATIVE MEASURE	ASSESSMENT
ENVIRONMENTAL IMPACT	Noise	Properties within Clapham benefit from removal of through traffic	No. properties experiencing: - Increase in noise 15 - Decrease in noise 316	301 properties experience net decrease in noise
CO_2 tonnes added 0 - 2000	Local air quality	Removal of through traffic by bypass will improve air quality within Clapham	No. properties experiencing: - improved air quality 400 - worse air quality 0	-409 PM_{10} -2549 NO_2
	Landscape	Bypass partially within local area of Great Landscape Value and would result in loss of pasture land.	-	Slight -ve
	Biodiversity	County Wildlife Site affected.	-	Slight -ve
	Heritage	No significant impacts.	-	Neutral
	Water	Even with mitigation, there may still be: a significant risk of polluting a sensitive watercourse and an aquifer used for public water supply during both construction and operation; and an impact on flood risk as the scheme is within a floodplain and bridges a river.	-	Moderate -ve
SAFETY	-	Bypass reduces pedestrian/vehicle conflict in village and replaces a section of poor standard single carriageway.	Accidents Deaths Serious Slight 311 9 94 359	PVB £10.5m 54% of PVC
ECONOMY	Journey times & VOCs	Faster journey times on new bypass.	peak inter-peak 4.6 mins 3.5 mins	PVB £27.6m 141% of PVC
	Cost	-	-	PVC £19.6m
	Reliability		Route stress Before 104% After 38%	Slight Low rel to PVC
	Regeneration		Serves regeneration priority area?	No
ACCESSIBILITY	Public transport	Will help to reduce peak journey times of existing local bus services .	-	Slight +ve
	Severance	Removes 80% of traffic from village.	-	Moderate +ve
	Pedestrians and others	Will improve accessibility for residents to local services.	-	Moderate +ve
INTEGRATION	-	Complements Bedford/Kempston package proposals and facilitates proposed Bedford North rail station with park and ride. Assists proposed local residential and commercial developments.	-	Positive
COBA			PVB £38.2m PVC £19.6m NPV £18.5m	BCR 1.95

Chapter 2

Strategic Transport Planning Evaluation – The Scandinavian Experience

Henning Lauridsen

Introduction

This chapter presents an overview of strategic transport planning and the evaluation methodology applied for this purpose in Scandinavia, that is Denmark, Norway and Sweden. There are similarities between the planning concepts applied in the three countries and two of them, Norway and Sweden, have developed their national transport planning systems along the same lines.

Various evaluation methodologies are applied in Scandinavia. In addition to *ex ante* evaluation methods applied as planning tools during the planning process, a considerable number of *ex post* evaluation studies have been carried out in Norway and Sweden. Due to these studies, Scandinavian experience also contributes to the picture of how planning processes and planning tools perform in the real world. The chapter, therefore, has a strong focus on *ex post* evaluations of strategic transport planning and the lessons learned from them. The concluding discussion of key strategic transport planning and evaluation issues in the last section is to a high extent based on experience gained through *ex post* evaluation studies.

The Changing Planning Concept

Current comprehensive and cross-sectoral strategic transport planning in Scandinavia has its conceptual roots in the national long-term planning of transport infrastructure projects, which started in the sixties. The planning concept has changed dramatically over the last three decades and we can discern three generations of national transport planning systems. The changes could be described as 'from project focus to strategies', but several other dimensions also changed.

The traditional Cost-benefit Analysis (CBA) method became the methodological basis for *the first generation of national transport planning systems* in Scandinavia. Government considered economic long-term planning highly important, and the transport agencies rapidly adapted the concept to their sector with the road agency as the leader. CBA was very well suited to address what became the predominant issue in road planning at that time, namely to identify those projects that should be implemented first. Long-term plans were very useful for the agencies as they provided a more stable framework for investments than the annual budget.

Second generation national planning systems for the transport sector were still characterised by a mode-by-mode approach and the road sector was clearly the leader. The planning concept, however, gradually became more complex than the socio-economic efficiency concept of the first generation systems.

Transport planning in the early seventies had to adapt to a more complicated world with several other dimensions than socio-economic efficiency. Environmental concerns and road safety came in at an early stage as important dimensions, which could not easily be integrated into the CBA methods. Gradually also regional development became an issue for transport planners. Urban transport planning had at that stage become an issue in itself, and new complex planning methods were developed for this purpose, among them much more sophisticated demand models. The planning concept changed accordingly and became more problem- and objective-oriented. Consequently, it focused much more on problem identification, definition of objectives aimed at solving the problems identified and alternative ways of solving them.

The resulting planning approach, therefore, typically included four phases: problem discussion and identification, definition of objectives, development of alternative sets of solutions for solving the problems identified and, finally, comparison of alternative solutions based on the objectives defined. The comparison provided a much better framework for the subsequent selection of the best solution by the decision-makers. The approach led to a clearer definition of the role of the planner in relation to the role of the decision-maker or politician. It also became a starting point for developing a range of impact assessment methods and multi-criteria analysis methods. These methods became important planning tools for the necessary evaluation of impacts of alternative solutions and for prioritising or selection of the best solution in a planning situation where several objectives should be met in the best possible way.

The division of labour between politicians and government ministries on one hand and the subordinated national agencies and enterprises on the other hand, gradually became an issue during the eighties. Liberalisation, privatisation and institutional reform were parts of the re-thinking of the principles for good governance in Scandinavia in the late eighties and early nineties. This led to the introduction of new objective oriented management principles. The principles were in line with the business administration principle of management by objectives. The new public sector management principles implied that politicians defined the objectives and made decisions on strategies, whereas the national agencies carried out their activities within the framework of the objectives set and the strategy defined.

The strong focus on objective-oriented management and planning for the transport sector led to more emphasis on policy matters by developing goals, objectives and strategies for the sector. All Scandinavian countries embarked on this in the nineties. Parallel to this, the principle of sustainable development was introduced as an overall goal for the transport sector. This principle is now generally accepted in Scandinavia as elsewhere in Europe. It may have contributed to the development towards third generation planning systems where all transport modes are seen together in the planning process. One mode may, therefore, be prioritised in relation to another because it better meets the objectives, of which some may derive from the goal of sustainable development.

The principle of sustainable development became important also for the development of the systems for environmental impact assessment (EIA) and strategic environmental assessment (SEA) in Scandinavia (Lerstang, 1997). Methods in this respect fall outside the scope of this chapter. They are, however, very important for transport planning.

The main features of the third *generation national transport planning system* as currently applied in Scandinavia are inclusion of all transport modes in the planning process, prioritisation of measures across sectors and use of alternative strategies to demonstrate the latitude for decision-making. Ideally such measures should include not only investment projects but also other measures found relevant, such as pricing and regulatory measures. The planning concept is based on the model that politicians are responsible for defining objectives and at the end of the process decide on a strategy. This implies that the planners, which include the national transport agencies for air, railway, road and sea transport, develop alternative strategies based on a combination of measures from all sectors and assess the impacts of each alternative with respect to the various objectives defined. This provides the input for the politicians' decision on strategy. When the strategy is defined, the national transport agencies start their planning of implementation, usually through ten-year action plans where implementation of the individual measures is prioritised in time and where the programmes for the first four years are usually more detailed.

Contents

The chapter describes the national strategic planning systems, which at this stage are more developed in Norway and Sweden than in Denmark. The *ex post* evaluation studies presented all relate to the former two countries. The chapter, consequently, has a stronger focus on Norway and Sweden than on Denmark, but there are indications that the planning system in Denmark may move in the same direction as in the other two countries.

The chapter makes an attempt to define the terminology applied in respect of planning and the corresponding *ex ante* evaluation. Three levels are identified, namely: the conceptual level, the approach level and, finally, the planning tools or evaluation methods level. A more detailed discussion of the contents of above three levels appears in a section below.

Subsequent to this introduction, the next section presents and defines the planning and evaluation typology used and gives an overview of application in the three Scandinavian countries. Sections three and four present the most important *ex post* evaluations of the national transport planning processes in Norway and Sweden. Finally, the main conclusions of the chapter are summarised.

The Current Evaluation Methodology

A wide range of evaluation approaches and methods is currently applied in strategic transport planning in Scandinavia. In addition to *ex ante* evaluation methods, recent experience includes a number of *ex post* evaluation studies. This section presents a review of typology and an overview of the current planning methodology as applied in each country.

Typology and Main Features of the Planning and Evaluation System

The chapter deals with evaluation in two different situations. The first is a planning situation where the various evaluation methods are applied as planning tools within a certain planning concept and in relation to a specific planning approach. The second is an *ex post* evaluation situation where the quality of the planning process or the various methods applied is assessed. The main focus of the chapter is, as stated in the introduction, the former situation, i.e. the application of an evaluation methodology in a planning situation, or more specifically for the purpose of strategic transport planning. There is, however, also a focus on *ex post* evaluation because a considerable range of *ex post* evaluations of strategic transport planning has been carried out in Scandinavia and because such evaluations are necessary if we want to gain knowledge about how evaluation methods perform in real life planning situations.

A planning system and the corresponding evaluation can be structured in a hierarchy of at least three different levels.

- the overall planning concept, i.e. the basic principles or ideology, which provide the conceptual framework for the planning system. The overall planning concept will usually be fixed in any given planning situation and, therefore, provide general guidelines for the approach level;
- the planning and evaluation approach, i.e. the methodological approach or system that determines the use of the individual evaluation methods and that results in the design of a planning process which tallies with the conceptual framework;
- the planning and evaluation tools, i.e. the individual methods or tools that are used in the various stages of the planning process.

The new third generation concept that Scandinavia currently considers the most relevant for strategic transport planning, is objective-oriented and cross-sectoral. This implies that planning is seen as a problem solving process that shall respond to sets of goals, objectives and criteria which to some extent may be contradictory, and aim at achieving them in the best possible way. The planning concept develops over time and can be considered a paradigm, which lasts for a certain period before it shifts.

Norway and Sweden have developed strategic transport planning systems that reflect the new concept, whereas the Danish planning system still partly reflects previous concepts. Due to this difference, the general methodological approach to strategic transport planning also varies considerably within Scandinavia. The approaches applied in Norway and Sweden appear to be more process-oriented and more based on system analysis than the approach used in Denmark. Elements of scenario techniques are also applied.

The differences between the three countries are fewer with respect to the specific evaluation methods, which are applied as planning tools within the various stages that are defined by the planning process and the corresponding overall methodological approach. A considerable range of evaluation methods and supporting planning tools are applied in all or most countries, among them cost-benefit analysis and multi-criteria analysis. Increasingly, evaluation tools are integrated with transport demand models, which provide important input for the assessment of impacts of the various measures considered in the planning process.

Below follows for each of the three countries, a brief description of the current transport planning systems and a review of the overall methodological approach to strategic transport planning and the evaluation methods applied.

Country Experience Denmark

Strategic transport planning in Denmark is still primarily based on a sectoral approach. The Government has, however, already developed most of the framework for an objective-oriented cross-sectoral system. The preparation of transport sector objectives and overall sector strategies (Trafikministeriet, 1993a) and the formulation of a national transport policy (Trafikministeriet, 1993b) were important steps in this direction. A recent report from the Transport Council outlines elements of a more comprehensive framework for an objective-oriented planning system, which in principle should be extended to include all transport modes (Transportrådet, 1999).

The Danish national road agency has developed an overall plan for investment in the national road network. The road agency has also developed a transport demand model that can be adapted to various elements of the network and applies cost-benefit analysis for evaluation of new projects (Leleur, 2000). The Danish railway track agency more recently also applies cost-benefit analysis for project evaluation. It seems that the CBA method applied by the agency is not completely compatible with the corresponding method applied by the road agency.

It, therefore, appears that the transport planning system currently applied in Denmark is more project-oriented and less strategy-oriented than in the other Scandinavian countries. This may, partly, be due to the difference in planning situation. The major strategic decision-making problems in Denmark have until recently been connected to the huge strait crossing projects across the Great Belt and Øresund, and a new huge fixed link to Germany across the Femer Belt is now on the planning agenda. Additionally, the Danish national road network (state roads) currently constitutes, and has also in the past decades constituted, a less dominant part of the transport system than elsewhere in Scandinavia. The need for and possibly also the resources available for developing complex strategic road planning systems may, therefore, not have been the same in Denmark as in the other two countries. It was strategic road planning that paved the way for strategic transport planning in Norway and Sweden and the methodological approach developed for the road sector was subsequently transferred to cross-sectoral transport planning.

Country Experience Norway

Current national strategic transport planning in Norway is based on a third generation concept. The planning is objective-oriented ('public management by objectives'). All transport modes and all national transport agencies are included in a joint planning process. Measures shall, in principle, be prioritised across the sectors. The planners shall develop alternative strategies by combining measures from all four sectors and also propose a recommended strategy. Subsequent to this, the politicians shall decide on a strategy for development of the sector and the strategy shall provide the framework for implementation of measures controlled by the national transport agencies.

The last planning exercise, National Transport Plan 2002–2011, started in 1998. The Ministry of Transport and Communications and the Ministry of Fishery issued guidelines for a joint planning process including the four national transport agencies responsible for sectors of respectively air, rail, road and sea transport. The planning guidelines emphasised close co-operation between the agencies and stated that the planning process should be open and transparent and involve the various stakeholders. They also indicated a planning process where the agencies should develop four alternative strategies including the various policy measures available to them: an environmental strategy, a transport safety strategy, a regional development strategy and a strategy for efficient traffic flow. In addition to the four optional strategies, the agencies should recommend a strategy. All strategies should be evaluated across the transport sectors and their impacts assessed with regards to a set of criteria defined in the guidelines (Sekretariatet for Nasjonal transportplan, 1999). The regional level was drawn into the early stages of the process and each county was requested to deliver a planning document on current problems and challenges as seen from their perspective (Stenstadvold and Lerstang, 1999).

The planning document prepared by the agencies should provide the framework for the subsequent strategic and comprehensive political decisions aimed at improving efficiency in the transport sector and strengthening the interplay between the transport modes.

The agencies delivered their planning report (Sekretariatet for Nasjonal transportplan) in September 1999 and the initial schedule for the political process envisaged that the ministries prepare a White Paper and that the Parliament discusses the issues and decides on a strategy during the first half of 2000. This schedule has been revised as the White Paper was published only in September 2000 (Samferdselsdepartementet, 2000). The White Paper reviews the issues of the planning report of the agencies and presents a recommended strategy for the consideration of the politicians. The strategy proposed by the White Paper is more detailed and specific than the one presented by the agencies, particularly with regards to the investment programme, which, for instance, mentions major investment projects in the various sectors and indicates financial allocations for trunk roads in the national corridors.

The next stage was the political process in Parliament in early 2001. Subsequent to this, the agencies are preparing more detailed action plans, i.e. implementation programmes for the period 2002–2011. The implementation programmes are more detailed for the first four years (2002–2005) than for the last six years. The White Paper specified that the four agencies should be responsible for preparing their own implementation programmes. The agencies should, however, co-ordinate their programme based on the outcome of the political process. The programmes should also be submitted to the regional level for comments. The National Transport Plan is a rolling planning system with four years intervals. The preprarations for the next planning round, which will cover the period 2006–2015, has therefore already started.

It appears from the above description, that the methodological approach to national strategic transport planning in Norway is highly process-oriented and it is based on system analysis. The methods applied for evaluation include cost-benefit analysis and multi-criteria analysis and the input to these methods is provided through a variety of other methods and tools, among them national transport demand models and various methods for impact analysis.

A number of *ex post* evaluations of the current process have been carried out or are planned. Such evaluations are carried out systematically to gain experience and to facilitate further improvement of the planning system. The following *ex post* evaluation studies related to the current planning process have been completed or are being carried out:

- Evaluation of the planning process at the regional level; completed and documented 1999 (Stenstadvold and Lerstang, 1999).
- Evaluation of the joint planning process of the national transport agencies; completed and documented 2000 (Ravlum, 2000).
- Evaluation of the political decision-making process in Parliament; ongoing.
- Evaluation of the implementation programming of the national transport agencies; ongoing.

In addition, a comparative study of the *ex post* evaluations of the planning processes in Norway and Sweden is being carried out. The second of the above evaluation studies is described in more detail in the third section below.

Country Experience Sweden

Sweden has substantial experience with national strategic transport planning and the current planning exercise represents a second round based on the same third generation concept as applied for the first round. The planning is objective-oriented and considerable efforts have recently been put into formulating the national transport policy and in making the objectives defined by Parliament operational in a planning context (Kommunikationsdepartementet, 1997). All transport modes and national transport agencies are included in a joint planning process. The measures controlled by the agencies shall, in principle, be prioritised across the sectors. The planners shall develop alternative strategies by combining measures from all four sectors. Subsequent to this, the politicians shall consider the optional strategies and, finally, decide on the strategy for development of the sector. This strategy shall provide the framework for implementation of measures controlled by the national transport agencies.

This implies a planning system that at the conceptual level is very similar to the Norwegian system as described in the section above. There are, however, important differences in the planning and evaluation approach. Another interesting difference is that much more efforts have been put into developing national and regional demand models and corresponding evaluation methods in Sweden than in Norway.

The strategic transport planning system in Sweden is divided into two phases. The first comprises an analysis of the current situation and the second is the so-called strategic analysis, which again comprises two different components. Accordingly, the present planning round started with the situation analysis, which was undertaken in 1998 by the Swedish Institute for Transport and Communications Analysis (SIKA) assisted by the four national transport agencies (SIKA, 1998). The report on the first phase also includes a review of the results of the previous planning round and a summary of various *ex post* evaluations made by transport agencies and others bodies involved in or having an interest in the planning process. The report recommended that a number of analyses of special strategic issues be included in the second phase.

The second phase was based on planning guidelines issued by the Government. The guidelines stated that the strategic analysis should investigate three different strategic development options or strategy alternatives: the Socio-economic (most efficient) Alternative, the Road Safety and Environment Alternative and the Regional Development Alternative. In addition, the planners should undertake analyses of twelve specific strategic issues. The results of the latter analyses should provide the framework for developing the three national alternatives.

The strategic analysis should be carried out jointly by the Swedish Institute for Transport and Communications Analysis (SIKA) and the four national Swedish transport agencies responsible for air, rail, road and sea transport. The analysis should provide the background for the Government's proposal to Parliament on strategies for development of transport infrastructure for the period 2002–2011.

At the same time, county authorities were directed by Government to develop regional development packages of transport measures aimed at promoting development of local and regional industries. These packages should again provide the building blocks for the above national strategy called 'The Regional Development Alternative'.

The agencies delivered their planning report in November 1999 (SAMPLAN, 1999). After that, the Swedish process appears to have halted, and the Government's proposal to Parliament on strategies for development of transport infrastructure is still pending. The political process has therefore not yet started.

Subsequent to the political process, the agencies should prepare more detailed implementation plans for the period 2002–2011. It is not clear at this stage how this process will proceed without the planned political guidance.

The Swedish transport plan is a rolling planning system with four years intervals. The development of the approach for the next planning round has commenced. The absence of political decisions, however, makes this more difficult.

The methodological approach to national strategic transport planning in Sweden is even more comprehensive than in Norway. It is clearly process-oriented and based on system analysis. The methods applied for evaluation include cost-benefit analysis and elements of multi-criteria analysis. The input to these analyses is provided through a variety of other methods and tools, among them national transport demand models, regional transport demand models and various methods for impact analysis. The methods are currently being streamlined and integrated. The output from the demand models will, for instance, in the future feed directly into the cost-benefit method.

There is a tradition in Sweden for carrying out systematic *ex post* evaluations. A number of such studies were carried out for the previous planning round. Recently, two *ex post* evaluation studies concerning the current planning round have been completed:

- Evaluation of the planning process at the national and regional level (Lauridsen and Ravlum, 2000a).
- Evaluation of the analyses of strategic issues and the decision-basis for measures (Larsen and Rekdal, 2000).

The above evaluation studies are described in more detail in the fourth section below.

An Overview

The table below summarises the description of the planning concepts and the different types of methodology currently applied in Denmark, Norway and Sweden for national strategic transport planning.

Table 2.1 Overview of Planning Concept Features, Evaluation Approaches and Methods for National Strategic Transport Planning in Scandinavia

Level	Item	Denmark	Norway	Sweden
1) Planning concept	Objective-oriented	x	x	x
	Strategy-oriented		x	x
	Project-oriented	x		
	Cross-sectoral		x	x
2) Approach	Process-oriented	(x)	x	x
	System analysis		x	x
3) Methods & supporting tools	Cost-benefit analysis	x	x	x
	Multi-criteria analysis	(x)	x	(x)
	Impact analysis	x	x	x
	National transport demand models		x	x
	Regional transport demand models			x
	Other demand models	x		

The table clearly shows the similarities between the planning and evaluation systems in Norway and Sweden and the somewhat different situation in Denmark. If we had included *ex post* evaluation, the differences would have become even stronger as *ex post* evaluation studies are currently applied systematically for learning by experience in Norway and Sweden but not in Denmark.

Ex Post Evaluation of National Planning Processes in Norway

This section is based on two *ex post* evaluation studies. The first dealt with the political process of the Norwegian Road and Road Traffic Plan 1998–2007. The second looked into the joint planning process of transport agencies concerning the National Transport Plan 2002–2011.

The Norwegian Road and Road Traffic Plan 1998–2007

The Norwegian Road and Road Traffic Plan for the period 1998–2007 was presented in a White Paper to Parliament in 1997. This was the first Norwegian example on a national transport policy document cum investment plan aimed at providing the background for a strategic decision-making process in Parliament. The document presented goals and objectives for the sector, described the measures required for achieving the objectives and described the results of calculations made for demonstrating the impacts of the various policy options or strategies. Based on this, the members of Parliament and, especially, the members of the Standing Committee on Transport and Communications, were expected to make more rational and intentional decisions. The strategies or policy options were presented as four alternative packages of investment projects and other measures, namely a strategy for more efficient traffic flow, a strategy for reducing negative environmental impacts, a strategy for improved traffic safety and, finally, one for rural development. The policy part of the White Paper was much more comprehensive and the investment plan was less detailed than in previous plans in order to give the decision-makers a better basis for policy discussions and prevent detailed discussions at project level.

An *ex post* evaluation study of the planning process was carried out subsequent to the political discussion in Parliament (Ravlum and Stenstadvold, 1997). The aim of the study was to examine whether the political process in Parliament was in accordance with the intention of a more goal and policy oriented process with less focus on individual projects, and whether the members of the Standing Committee found the White Paper adequate and useful as a basis for policy decision.

The Process should be Policy-oriented and Cross-sectoral The study showed that the politicians generally were satisfied with a less detailed plan. The majority would prefer a more policy oriented and even less detailed investment plan as background for the political discussion and decision. This was, however, subject to the condition that politicians had real influence over the policy framework and that there was a clear linkage between policy decisions made at the political level and the subsequent implementation by the road agency. The politicians further felt that a more comprehensive policy framework should include all modes of transport (air, rail, road and sea transport) as well as public transport. Such framework would facilitate a policy-oriented process by widening the latitude for political decisions and facilitate prioritising among transport modes. It would also make more measures available for a comprehensive political assessment.

The politicians emphasised the need for an institutional framework that could support a comprehensive policy approach. This would include a better division of labour between the national authorities and the regional authorities, thereby securing a more optimal resource allocation between the modes of transport. Further, such framework would secure that the needs for public transport were taken better into account, whereby public transport to a higher decree could become an alternative solution to car transport.

The Process should be more Objective-oriented The ideal model for rational planning and decision-making includes several elements. It starts with the definition of goals and objectives based on the problems at hand. This is followed by a search for the most effective measures to achieve the objectives and finally a selection of those actions, which within available financial resources most efficiently fulfil the desired ends. In this sense the White Paper on the National Road and Road Traffic Plan was not conducive to a rational political decision-making process in the Standing Committee.

This was mainly due to the presentation in the White Paper, which described the alternative transport strategies partly in isolation from the proposed policy measures and the investment plan. This made it difficult to see the internal logic from the strategies to the various measures and actions selected for inclusion in the plan. It appeared that the politicians, therefore, decided on the components of the investment plan and made policy recommendations without taking the strategies into account. They were rational policy-makers in the sense that their decisions reflected the lowest possible cost. The strategies presented in the plan were, therefore, not used as intended but became more superficial elements in the ensuing hectic decision-making process.

Other factors also contributed to marginalising the alternative strategies in the political process. The White Paper was presented to Parliament only weeks before an election campaign began. This contributed to the fact that the politicians took party-interests more into account than otherwise and, therefore, put less emphasis on the new approach. Further, the Standing Committee had only a few weeks to consider and discuss the White Paper. This did not make it easier for the politicians to familiarise themselves with the new planning approach.

Main Conclusions Two main conclusions emerge from the *ex post* evaluation of the political decision-making process as described above. The first is that the White Paper on the national road and road traffic plan was not adequate in providing a sufficient basis for a more rational political decision-making process. That is a process, which includes selection of those measures that most effectively meet the objectives defined and the actions that most efficiently fulfil the desired ends. On the other hand, the plan did encourage a less detail-oriented approach, which, however, was not considered comprehensive enough by the politicians.

The other main conclusion is that other factors, beyond the contents of the plan, also contributed to a less rational planning process than intended. Two factors, in particular, were important in this respect. The plan was presented only weeks before the beginning of an election campaign and, secondly, there were tight overall time constraints for the political process.

The National Transport Plan 2002–2011

The four national transport agencies responsible for air, road, rail and sea transport in Norway were in 1998 directed by the Ministry of Transport and Communications and the Ministry of Fishery to prepare a joint proposal for a long-term national transport plan for the period 2002–2011. An *ex post* evaluation study of the joint strategic planning process was carried out in 2000 (Ravlum, 2000). Below follows a presentation of the study and its main conclusions.

Background and Planning Approach The planning guidelines from the two ministries emphasised the need for close co-operation between the agencies throughout the planning process and that the planning process should be open and transparent and involve the various stakeholders. The guidelines stated that the planning document should provide the background for strategic and comprehensive political decisions aimed at improving efficiency in the transport sector and strengthening the interplay between the transport modes.

The planning approach was specified in the guidelines. The agencies should develop four alternative strategies taking into account the various policy measures available to them: an environmental strategy, a road safety strategy, a rural development strategy and a strategy aimed at more efficient traffic flow. In addition, the agencies should prepare a recommended strategy. All strategies should be evaluated across the transport sectors and their impacts assessed in respect of the criteria defined in the guidelines.

Main Study Areas The *ex post* evaluation study of the strategic planning process looked into the following three main questions:

- To what extent was the process characterised by real co-operation between the transport agencies (including evaluation of measures and effectiveness across sectors), transparency and involvement of the various stakeholders?
- Which factors contributed to the actual planning process?
- Which factors could contribute further to real co-operation between the agencies and cross-sectoral effectiveness?

The evaluation of the planning process was based on semi-structured interviews of 25 interviewees from the four transport agencies and the two ministries involved.

The Agencies did Co-operate to a Certain Extent The four transport agencies did work together to establish a common understanding of the problems, to prepare a common description of conditions in the transport sector and to generate the four alternative strategies. The recommended strategy, however, was developed sector by sector and separately in each agency without involvement of other agencies. With the exception of the more general policy statements, the joint recommended strategy became a product of four sector plans rather than a product of a joint process.

The study concluded that the actual co-operation primarily was one of presenting general transport policy principles, rather than mutual scrutinising of each agency's measures and means. On the other hand, the alternative strategies, which illustrate how policy measures could be combined in different ways, were worked out through discussions between the railway track agency and the road agency.

The road agency integrated the planning process into its normal planning procedures, involving both corporate management and the organisation as such. Also the railway track agency involved its organisation in the planning process, although it had less capacity in respect of human resources than the road agency. The coastal transport agency was eager to participate in the planning process, but had less experience in strategic planning and less capacity than the other agencies. The civil aviation agency had fewer interests in common with the other three agencies and did not prioritise the work as much as the latter.

Limited Cross-sectoral Evaluation The agencies agreed at an early stage that they did not have the necessary quantitative methods and planning tools to assess the effectiveness of their measures across the transport sectors. They did, however, not agree on using more qualitative methods for cross-sectoral evaluation due to a disagreement on the extent of competition among the transport modes.

The agencies were not able to evaluate impacts in a comparable way. The coastal and civil aviation agencies were not able to assess impacts at all. They faced, additionally, problems in handling intermodal transport in a satisfactory way. Consequently, the planning process did not come very far in assessing gains from co-ordination and overall effectiveness across transport modes. Several interviewees claimed that the agencies could have done more in this respect. The agencies felt, however, that it would be difficult to agree on such an overall assessment due to lack of appropriate methods and a shared knowledge basis.

More Transparency and More Involvement of some Stakeholders Stakeholders that usually are not in close touch with the transport agencies and the planning process felt they had gained more influence through participation in an external reference group. On the other hand, non-transport government agencies that also participated in the reference group found that involvement, at the same level as the NGOs, reduced their formal position as government agencies. Regional and local authorities and politicians found the process more transparent and participatory than before, even though the regional input did not have any significant influence on the resulting national priorities.

Institutional Differences Cause Conflicting Interests and Unequal Power The road agency's previous experience with this type of strategic planning gave it the strongest position. Budget size, human resources and many interests at stake were all factors that contributed to strengthening this agency's position as primus inter pares in the process. The rail agency was the second strongest agency. Most discussions took place between these two agencies, which also carried out most of the planning work. The coastal agency scored lowest on the institutional variables, even though it was the civil aviation agency that participated least in the co-operative process. The latter agency is self-financing through user fees and less dependence on government funding and a relatively autonomous status may explain its low level of participation.

It appears that the agencies to a great extent behaved in accordance with their institutional characteristics. The civil aviation agency had least interests at stake in the process and was least involved. The coastal agency had limited capacity and resources and could, therefore, not participate as actively as it wanted. Most discussions took place between the strongest agencies, namely the road agency and the rail agency, with the former as the lead partner.

The Planning Guidelines may have Amplified Institutional Differences The planning approach of the guidelines did not suit the civil aviation and the coastal agencies too well. Neither of them was able to assess the impacts of their own measures or differentiate their means into alternative strategies. The guidelines assigned the formal leading and co-ordinating role and the responsibility as planning secretariat to the road agency. This might have had a positive as well as a negative effect on the agency's ability to promote its own interests. The leading role gave the agency an opportunity to put its stamp on the work at an early stage. On the other hand, this role also entailed responsibility for promoting joint interests. Disagreements among the agencies, for instance on definitions and on what should be included in the joint planning process, strengthened the civil aviation agency's initial reserved attitude to the process.

Most interviewees felt the recommended plan should have included measures beyond the control of the four agencies. The agencies described such measures in the more general part of the planning document and were not given unlimited opportunities to base the recommended strategy on policy conditions not already approved by the ministries or Parliament. The transport policy objectives were, on the other hand, quite ambitious and broadly defined by the ministries. This created a gap between objectives and measures available to achieve them.

Factors that might Improve the Planning Process The transport policy objectives that the agencies were asked to achieve, should have been more in line with the measures the agencies actually control. Moreover, too widely defined and partly inconsistent transport policy objectives may draw attention away from measures controlled by the agencies towards policy measures controlled by others.

In spite of the short time available, the agencies could have put more emphasis on co-ordinating their recommended strategies and not only the more general policy aspects. They did not scrutinise each other's input. The agencies could have put more emphasis on developing a joint recommended strategy and not merely presented an aggregate of four separate ones. The four alternative strategies should also have been used more directly for developing the recommended strategy.

There was a mismatch between strategic planning and the agencies' tradition of project focus, which led to a bottom-up rather than a strategic top-down approach. A division of the planning process into a first phase of general policy principles and a second phase of developing alternative and recommended strategies might be a better solution. This could also pave the way for more relevant involvement of the regional level at an appropriate time in the planning process.

Before the agencies started the planning process, a clearer distinction should have been made between efforts that must be carried out jointly and matters that do not necessarily involve all agencies. A better scheduling of the process could facilitate closer participation of the regional level. The common knowledge basis, especially compatible evaluation methods, needed improvement. Such improvement should take place before the next revision of the plan.

A joint a secretariat for the National Transport Plan process that includes staff from all agencies should be considered. Such secretariat should consist of employees holding key positions in the agencies to ensure ownership and agency responsibility for the cross-agency operations.

Ex Post Evaluation of the National Transport Planning Process in Sweden

The Swedish Institute for Transport and Communications Analysis (SIKA) and the four national transport agencies responsible for roads, railway tracks, civil aviation and sea transport were by Government Decision of 1999 directed to carry out jointly a national strategic analysis. An *ex post* evaluation of the planning process was carried out late in 1999 (Lauridsen and Ravlum, 2000a).

Background and Objectives of the Study

The government guidelines stated that the strategic analysis should investigate three different strategic development options (strategic alternatives): a socio-economic (most efficient) alternative, a road safety and environment alternative and a regional development alternative. In addition, analyses of twelve specific strategic areas should be carried out. The results of the latter analyses should provide parts of the background for development of the three national strategic development options. The guidelines also stated that the analysis should provide the background for the Government's proposal to Parliament on strategies for development of transport infrastructure for the period 2002–2011.

County authorities were simultaneously directed to develop regional development packages of transport measures aimed at promoting development of local and regional industries. These packages should again provide the building blocks for the above regional development alternative.

The overall aim of the evaluation study was to identify and explain which factors contributed to and characterised the actual planning process.

The objectives defined three main areas of investigation:

- Was the planning process organised in such way that it promoted comprehensive planning and cross-sectoral assessment, and did the contents of the process facilitate achievement of the goals and objectives of the national transport policy?

- To which degree were the results of the analyses of the 12 specific strategic areas used when developing the three national strategic options and designing the measures included in these?
- To which degree were the policy differences between the three strategic options reflected in the measures selected, and how was socio-economic efficiency taken into account when designing measures?

The evaluation was based on semi-structured interviews of selected central and regional participants in the planning process. The interviewees were key planners in the five national bodies responsible for the national plan and in two counties selected for the purpose. The study investigated how these participants perceived the process.

Organisation of the Planning Process and Comprehensiveness

The assessment of the comprehensiveness of the national planning process in Sweden was related to four possible and increasingly ambitious levels for comprehensive and cross-sectoral planning: (1) participation of all transport sectors in the planning process, (2) application of cross-sectoral evaluations, (3) cross-sectoral prioritisation, and (4) application of a system approach including both investments and other measures.

Work at the Central Level

All four national transport authorities participated. Two of them, the civil aviation and the sea transport agencies were, however, only lightly involved and mainly focused on ensuring that their areas of responsibility were treated correctly in the planning report. The planning process at the central level was neither perceived comprehensive nor cross-sectoral. The work was characterised by bilateral discussions, mainly between the rail agency and SIKA and the road agency and SIKA. All transport agencies felt that SIKA had the main responsibility for the planning process. SIKA was also predominant in respect of report writing.

Application of transport demand models and socio-economic analyses should in principle enable evaluation of measures across the transport modes against each other. The demand models were, however, not fully developed and operational as scheduled and socio-economic analyses of investment projects could therefore not be completed within the time limit. No corridor analyses were carried out.

It was intended to use marginal cost principles to establish the financial frameworks for railway investment projects and road investment projects respectively. Due to the delay in model development, such calculations could not be undertaken. The overall allocation to rail and road investments was instead based on a negotiation approach, which led to a result very close to the previous allocations to the two modes. Cross-sectoral prioritisation among modes was therefore not based on the intended analytical approach.

A system approach including both investments and other measures was only applied to a rather limited degree. The resulting planning report includes various general cross-sectoral analyses. The national analysis and the development of the three strategic options were, however, mainly limited to projects and measures that are controlled by the transport authorities. Subsidies to public transport as a measure, which can reduce the need for investment in infrastructure projects, was not considered when designing the three strategic development options. This measure was, however, discussed in general for urban areas in one of the analyses of specific strategic areas.

Work at the Regional Level

The planning process at the regional level had a somewhat different character. It was more geographically specific and more problem-oriented. The regional level looked to some extent into transport corridors and major projects. It appeared as a somewhat more comprehensive process than the process at the national level but it was less analytical and less focused on strategic considerations.

Use of Results from the Analyses of the Twelve Specific Strategic Areas

The analyses of strategic areas had to a varying degree an impact on the actual development of national strategic options. At the regional level, knowledge about the results was very limited.

Four analyses, which concerned development scenarios for society, policy on CO_2 emissions and transport demand for passenger and freight traffic, were particularly important as they provided a common basis for the three national strategic options. The work was carried out jointly by the transport authorities, and the results were agreed. Disagreement was limited to the use of the results, notably whether a new CO_2 levy on fuel should be introduced.

In the analysis of road safety, the application of a CO_2 levy and speed limitations instead of other and less controversial measures was discussed intensely, but a common agreement in this respect was not reached. This analysis became, however, highly important for the design of the road safety measures used in at least two of the strategic options.

Two railway related analyses were carried out by the rail agency and other agencies were only marginally involved. The results were not commonly agreed. The analysis was important for the railways, but it had no clear impact on the development of the three strategic options. All parties were satisfied with the analysis of port structure and sea transport. It had, however, no notable impact on the development of the strategic options.

The analysis on maintenance of road and railway infrastructure had a strong impact on the development and design of two of the strategic options. There was, however, a common understanding that the results concerning the optimal road maintenance level were tentative and that further research in this respect was needed.

It appears that the various analyses to a varying degree met the requirements of the planning guidelines. This reflects to some extent the limited time available for the analyses (Larsen and Rekdal, 2000). The analyses represented a new and useful element in the Swedish planning system, and there appears to be a need for more focus and professional emphasis on them in the future. Some are highly critical for prioritisation among the various categories of measures in the final strategic plan, in particular the balance between investment and maintenance of infrastructure.

Were Policy Differences Reflected in Measures selected and was Socio-economic Efficiency Taken into Account?

The CO_2 levy and speed limitations were combined with use of shadow prices for the road safety and environment option. Measures selected for this option, therefore, differed from measures selected for the socio-economic option.

The transport agencies did not explicitly select investment projects that primarily contributed to the achievement of the road safety and environment goals. The investment projects selected for the road safety and environment option were mainly the same as those selected for the socio-economic option. The regional development option differed more and included a wider range of projects. These projects were, however, not selected in such way that they clearly aimed at achieving the regional development goals. The counties did to a high extent give priority to projects that also were prioritised in the previous strategic planning round, although the regional development goals were quite different at that stage. It does, therefore, not appear that the most efficient measures for goal achievement generally were selected.

The study looked specifically into design of investment projects included in the three options. The project design was fixed and totally independent of the option. There were, however, variations among options for some other measures. The size of the packages of other road safety measures was much larger for the road safety and environment option, and there was less emphasis on infrastructure maintenance in the regional development option than in the other options.

Due to delays in transport model development, the socio-economic analyses of investment projects were not carried out as planned at the central level. Socio-economic analyses were not used at the regional level.

It appears, in conclusion, that transport agencies and counties ended up introducing investment projects that they previously had identified and, which they probably would have given high priority to, also under other circumstances. If we exclude the CO_2 levy and speed limitations in the road safety and environment option, differences among options mainly comprise the size of the various packages of measures and to some extent a different order of priority for investment projects. The study was, therefore, not far from concluding that: strategic options arrive and pass by, whereas investment projects survive.

Which Factors Contributed to the Planning Process?

A number of different categories of variables may have an impact on the planning process and the final plan. The most important are: the guidelines for the planning process, the planning agencies' interpretation of the guidelines, the knowledge basis and the way it is interpreted by the planners, the organisation of the planning process and, finally, the institutional characteristics of the participants.

Planning Guidelines The general Swedish guidelines for strategic transport planning, which form part of the governments transport policy (Kommunikationsdepartementet, 1997) did, obviously, have a major impact on the actual planning process. Both the general guidelines and the government guidelines of 1999 were, however, open to various interpretations, and the transport agencies expressed different views on key elements during the process.

Interpretation of the Guidelines The national transport agencies and SIKA had extensive discussions on the interpretation of the goals and objectives for the transport sector, and the relationship between operational objectives and overall goals, which have not yet been expressed in operational terms. There was also disagreement on how to apply the principle of socio-economic efficiency in the development of the three strategic options. The disagreement was reflected in the discussion on which measures to include in the various strategic options. This applied in particular to measures aimed at achieving sustainable development and road safety.

Both the rail agency and the road agency issued formal written statements of disagreement concerning the design of the road safety and environment option. Some of the interviewees felt that the process rather should have addressed problems than focus on whether solutions are economically beneficial. The disagreement concerning goal interpretation continued throughout the work and was hardly beneficial for the planning process or for smooth co-operation among the parties. In particular, the railway participants felt that their opinions and interests were disregarded.

The Knowledge Basis In parallel with the analyses of specific strategic areas, an extensive work on development of transport demand models linked to tools for impact analysis went ahead. The results of the two parallel activities should provide a strong professional knowledge basis for the development of the three strategic options. The model development was, however, not completed in time for active use in this respect.

A number of interviewees emphasised that model development work diverted the focus from other planning approaches, which would have been more relevant for the strategic analysis. It is a paradox that economic analyses of investment projects were emphasised much more than the analysis of maintenance need, which revealed serious methodological shortcomings and which provided the basis for the conclusions related to the far more important package of maintenance measures.

Organisation of the Planning Process The process was carried out in a short period of time compared to the high professional ambitions. Several of the analyses of strategic areas were completed only just before the completion of the final planning report, and most were carried out in parallel to the development of the strategic options. The development of the models was not completed before publishing of the final report. It is likely that more time would have made it easier to reach agreements on the various controversial matters that arose in the process.

All parties involved perceive the process and the results as 'owned' by SIKA. SIKA was the driving force in the process, wrote the report and, partially assisted by the Ministry, acted as a referee with respect to the various disagreements along the way. SIKA and the national transport agencies seem to have chosen somewhat different approaches and roles in the process.

The civil aviation agency and the sea transport agency were in several respects on the sideline. The two agencies found other decision-making arenas like the regional project planning process more important.

The rail agency held a position of defence. It seldom took the initiative and felt that its legitimate points of views were not taken duly into account. The agency was not able to deliver all the contributions requested. The involvement of corporate management was low. The agency had serious problems with the design of the strategic analysis and also with the transport models.

The road agency participated in a formally correct manner, delivering the contributions it was asked to make and having fewer conflicts with SIKA. Corporate management was more involved than that of the rail agency. The road agency had some problems in respect of the design of the national strategic options, particularly the safety and environment alternative.

SIKA saw its own role as one of scrutinising the policies and priorities of the four transport agencies. Accordingly, such role would easily lead to conflicts with the interests of the other participants.

Institutional Characteristics of the Participants The study could not fully explain the different ways the participants acted. Some aspects of the institutional and organisational set-up might, however, help explain how the various agencies interpreted their role.

Different economic and financial frameworks were important. The civil aviation agency and the sea transport agency are self-financed through user charges. Consequently, their investment projects or other activities were not as dependent on the strategic planning process as those of the other two agencies. In this sense, their interests were not really at stake.

The level of competence varied among the agencies. The rail agency and its employees have less experience of strategic planning. The agency had also experienced a high turnover of planning staff while that of the road agency had been stable.

Different organisational structures and cultures were at play. The road agency is a larger organisation and the whole organisation is more focused on planning than the rail agency. The road agency has a tradition of top management involvement in national planning. This might have given it more power in negotiations.

Different perceptions of the political environment are also possible explanatory factors. The road agency felt that politicians were somewhat critical towards roads. The strategic analysis might, therefore, have been seen as an appropriate arena for defending the core interests of the agency and it would be rational to put emphasis on the analysis. The rail agency seemed to be more comfortable with the politicians' attitudes to railways and the increasing sector budgets. The agency may therefore, have considered the strategic analysis less crucial for its interests.

The Planning Process at the Regional Level

The responsibility for the regional planning process was vested in the county authorities. The other participants looked upon themselves as contributors, emphasising a good relationship with the planners in charge. All participants, however, stressed their ownership of the process and its results. The regional offices of the national road and rail agencies participated most actively in the process. The civil aviation and the sea transport agencies were not actively involved, but the regional planners were nevertheless preoccupied with the role of these transport modes in the planning process.

The regional process appeared to be harmonious without serious disputes. The local planning authorities defined the objectives of the regional development alternative in a way that reflected a common understanding of the regional transport challenges among the participants. The two counties studied had relatively generous financial frameworks for their planning. They had, consequently, not the same need for strict prioritising as at the national level. It appeared that the regional planners mostly did what they were used to. Although the goals for the regional development alternative had changed since the last planning round four years earlier, the contents of the regional plans did not change substantially.

Conclusions

The review of the Scandinavian experience of national strategic transport planning shows that the current planning concept in many respects is fairly advanced and well designed. A range of *ex post* evaluation studies has been carried out and has given insight into current processes and methodologies. One of the major findings is that the institutional characteristics of the planning agencies play a very important role in the planning process.

Generally the current planning systems and planning approach in Norway and Sweden are similar and well in line with the requirements of the conceptual framework as developed over the last decades in Scandinavia. Denmark has a somewhat different and more project oriented approach. The conclusions below therefore primarily refer to Norway and Sweden.

The Relevance of the Planning Concept and the Methodological Approach

The Political Process Perspective

The relationship between planner and politician is a major concern for the planning approach applied in Scandinavia. The current planning system can only function effectively if politicians receive the information needed at the right point in time. It is difficult at this stage to assess the relevance of the current strategic planning process in respect of the subsequent political decision-making process in Parliament. It is, however, only when the technical planning process can be seen in the light of the political process, that we get full feedback about the appropriateness of the technical process.

A study about Parliament's discussion of the Norwegian Road- and Road Traffic Plan for the previous planning period casts some light on the matter. The study concluded that very few politicians did seriously consider and use the information on alternative strategies. It is, therefore, at this point in time not clear whether the strategic transport planning processes now applied really provide the most relevant decision information for the politicians. An ongoing study about the political process concerning the current National Transport Plan in Norway is, however, looking further into the matter.

The Planning and Methodological Approach

Generally the current approach is well in line with the requirements of the conceptual framework. It is objective-oriented, strategy-oriented and cross-sectoral. There are, however, some matters of concern.

One concern is the objectives that guide the planning process. There are cases where there are discrepancies between the objectives and the measures available to the planning agencies. A key conclusion is that policy objectives should be realistic and achievable within a reasonable time perspective compared to the range of measures available to the strategic transport planners. If the objectives are more ambitious and broadly defined, planners should also be allowed to use a wider set of measures, but that may shift focus away from the measures controlled by them.

Another concern is the strategy concept applied in Norway and Sweden. The planning agencies are requested to develop alternative strategies that include different combinations of actions and measures. The basic idea behind the strategy concept is that politicians discuss strategies rather than projects and, finally, select a strategy, which then provide the framework for the more detailed planning and implementation. Experience from *ex post* evaluation shows that there are good reasons to discuss whether the above approach to strategy-orientation is the best. It appears that the strategies developed by the planning agencies are not necessarily significantly different in respect of projects and measures included. Further, the little we know at this stage about the political process does not indicate that politicians find the strategies particularly useful in their decision-making process. The strategies may, however, be useful for others such as the Ministry of Transport.

The evaluation studies revealed serious problems, in respect of cross-sectoral prioritisation in both Norway and Sweden. It is still an open question to which extent these problems were due to insufficient knowledge or to lack of tools, which normally would be the conclusion of planners, or to which extent, the professional level of ambition for such planning is realistic. It seems, however that the planning approach and the ambitious processes were more to blame than the various evaluation methods and planning tools applied.

There are reasons to believe that the planning approach can be improved by re-organising the process. A division into a first phase of clarification of general policy principles through a set of analyses of strategic issues and a second phase of developing alternative strategies may be a better approach. Further, it appears that the regional level should play a bigger role in the national planning process. A two-phased approach would facilitate involvement from the regional level at an appropriate time in the first phase of the process.

Calculations of impacts of the three strategic options applied in Sweden revealed that differences among the alternative strategies were rather small at the national level. The same seemed to be the case in Norway. Generally, it appeared that ordinary investment projects led to small differences in overall impacts. Packages of specific physical measures, for instance road safety measures applied generally all over the country led to more significant differences. Application of 'global measures' such as introduction of a national CO_2 levy and lower general speed limits leads led to considerable differences. It is important to note that some strong measures, such as a national CO_2 levy is not controlled by the transport sector but falls under the Ministry of Finance.

Experience from both Norway and Sweden shows that the inherent institutional characteristics of the planning agencies are key factors in the process. The road agencies and the rail agencies played the most important roles, whereas the civil aviation agencies, which are financed through user charges, kept a low profile. It may, therefore, be worthwhile to consider if the process can be organised differently with a clearer distinction between efforts that must be carried out jointly and matters that do not necessarily involve all agencies.

The planning process at the regional level in Sweden had a different character than the process at the national level and appeared to function smoothly. It was more geographically specific and more problem-oriented. The regional level looked to some extent into transport corridors and major projects. It appeared a somewhat more comprehensive and cross-sectoral process than the process at the national level. The regional process was, however, less analytical and less focused on strategic considerations.

In Norway, a recent *ex post* evaluation study shows that regional and local authorities primarily were involved in the planning process at an early stage through their work on local challenge documents (Stenstadvold and Lerstang, 1999). Counties and local communities also at a later stage commented on the national transport plan proposal, primarily focusing on the county-specific priorities. Representatives of the counties and communities found that the actual planning process improved transparency and local political involvement compared to previous processes, even if shortage of time made hearings and political processes at the local level difficult.

The above experience gives reasons to raise the question whether in the future more of the planning process, or at least the preparatory phases, could be handled at the regional level. Concerns in respect of more responsibility to the regional level in the planning process were raised in the Norwegian study referred to above. A preliminary conclusion is that experience so far indicates a need for regional participation in national strategic transport planning and that such participation has functioned well in Sweden. The possibilities for regional participation in the future should, therefore, be investigated further.

In addition to the issues discussed above, it is worth mentioning that the use of *ex post* evaluation in Scandinavia represents a quality in itself. The planning process can only be said to be fully complete when *ex post* evaluations systematically are included as a last component. The Scandinavian *ex post* evaluation experience is focused on the planning process and the methods applied. *Ex post* evaluation is, however also required to establish whether the planned results are reached or, more specifically, to which extent the objectives are met and the expected output achieved. Such knowledge is in fact fundamental for an objective-oriented planning system. Experience shows than even the most sophisticated planning systems do not solve all problems. *Ex post* evaluation can, however, by identifying how they actually perform, contribute to the gradual improvement of such systems.

Further Development of Evaluation Methodologies

Ex post evaluations of planning methods in Sweden addressed the methods for cost-benefit analysis. The general conclusion was that the methods as such appear to be sound and sufficient for the purpose. Questions raised concerned primarily current practises. A more serious question, however, concerned the quality of input data, in particular traffic data (Larsen and Rekdal, 2000). It may, therefore, at this stage be more important to improve traffic data and to introduce a reliable system for documentation of such data, than to improve the methods for cost-benefit analysis. A previous evaluation study raised some doubt about application of the methods and asked if they had been adapted to the results wanted.

The evaluation of the Norwegian strategic planning process showed that the agencies were not able to evaluate impacts across sectors in a comparable way. The coastal agency and the civil aviation agency were hardly able to assess the impact of their own measures at all. In addition, the agencies faced problems in handling intermodal transport in a satisfactory way. There is consequently a need for developing compatible methods for all sectors.

There were shortcomings in respect of better methods for analyses of the specific strategic areas, which formed a new and important element of the Swedish strategic analysis. This applied not least to the analysis of maintenance needs. It is therefore important to improve methods for analysis of strategic issues.

There is obviously a need for further development of the evaluation methods and in particular some of the supporting tools such as the national and regional transport demand models. Current practises may, however, be a bigger problem than the methods per se and it appears that it is highly important to ensure that input data are of sufficient quality and documented properly. Development of the specific evaluation methods must be done in such way that they fit into the overall evaluation approach.

References

Kommunikationsdepartementet (1997), *Transportpolitik för en hållbar utveckling (Transport Policy for a Sustainable Development)*, Prop. 1997/98, p.56, Kommunikationsdepartementet, Stockholm.

Larsen, O.I. and Rekdal, J. (2000), *Evaluering av den svenske innretningsplanleggingen. Strategiske analyser og beslutningsunderlaget for tiltak (Evaluation of Strategic Transport Planning in Sweden, The Strategic Analyses and the Decision Basis for Measures Applied)*, TØI Working Report 1154/2000, Institute of Transport Economics, Oslo.

Lauridsen, H. *et al* (1995), *Strategisk vegplanlegging. Oppsummering av erfaringer med strategisk planlegging (Strategic Road Planning)*. TØI Report 316/1995, Institute of Transport Economics, Oslo.

Lauridsen, H. (2000), *Strategic Transport Planning and Evaluation, The Scandinavian Experience*, Paper presented to the Second TRANS-TALK Workshop, Brussels, November, 2000, TØI Working Report 1177/2000, Institute of Transport Economics, Oslo.

Lauridsen, H. and Ravlum, I.A. (2000a), *Evaluering av prosessen i den svenske innretningsplanleggingen (Evaluation of the Strategic Transport Planning Process in Sweden)*, TØI Report 469/2000, Institute of Transport Economics, Oslo.

Lauridsen, H. and Ravlum, I.A. (2000b), *Strategisk transportplanlegging i Norge og Sverige – Evaluering av planprosessen (Strategic Transport Planning in Norway and Sweden – Ex post Evaluation of the Planning Process)*, Paper presented at the Transport Conference at Aalborg University 2000, Proceedings, Vol. 2, The Danish Transport Council and the Transport Research Group at Aalborg University, Aalborg.

Leleur, S. (2000), *Road Infrastructure Planning, A Decision-oriented Approach*, Polyteknisk Forlag, Lyngby-Copenhagen.

Lerstang, T. (1997), *Which Strategic Planning? The Emerging Interest for Application of Strategic Environment Assessment in Transport Planning at Various Levels*, in the Proceedings from the 14th International Symposium on Theory and Practice in Transport Economics, Which Changes for Transport in the Next Century? Innsbruck 21–23 October, 1997, ECMT.

Norges offentlige utredninger (1989), *En bedre organisert stat* (*A better organised State*), NOU 1989, p.5., Forvaltningstjenestene Statens Trykkingkontor, Oslo.

Norges offentlige utredninger (1993), *Nytt overordnet styringssystem for Statens vegvesen (New Governance System for the National Public Roads Authority)*, NOU 1993, p.23, Statens Forvaltningstjeneste, Seksjon Statens Trykning, Oslo.

Ravlum, I.A. (2000), *Helhetlig, tverrsektorielt og åpent? – Transportetatenes samarbeid om Nasjonal transportplan 2002–2011 (Comprehensive, Cross-sectoral and Transparent? – The Joint Planning Process of the National Transport Agencies for the National Transport Plan 2002–2011)*, TØI Report 488/2000, Institute of Transport Economics, Oslo.

Ravlum, I.A. and Stenstadvold M. (1997), *Fra vegstubber til strategi og helhet? Stortingets behandling av norsk veg- og vegtrafikkplan 1998 – 2007 (From Stumps of Road to Strategy and Comprehensiveness, Parliament's Discussion of the Norwegian Road- and Road Traffic Plan 1998–2007)*, TØI Report 374/1997, Institute of Transport Economics, Oslo.

Riksrevisionverket (1997), *Vägverkets, Banverkets og länens förslag til infrastrukturinvesteringar åren 1998–2007 – En kvalitetsbedömning av beslutsunderlaget (The Road* Authority's, *the Railway Track* Authority's *and the Counties' Proposal on Infrastructure Investments 1998–2007 – A Quality Assessment of the Decision Basis)*, Report RRV 1997, p. 60, Riksrevisionsverket, Stockholm.

Samferdselsdepartementet (2000), *Nasjonal transportplan 2002–2011 (National Transport Plan 2002–2011)*, St.meld. nr. 46 (1999–2000), Samferdselsdepartementet, Oslo.

SAMPLAN (1999), *Strategisk Analys – Sluttrapportering av regeringsuppdrag om inriktningen av infrastrukturplaneringen för perioden 2002–2011 (Strategic Analysis – Final Report on the Policy for Infrastructure Planning for the Period 2002–2011)*, SAMPLAN Report, p. 2, SAMPLAN, Stockholm.

Sekretariatet for Nasjonal transportplan (1999), *Forslag til Nasjonal transportplan 2002–2011 (Analyses and Recommendations on the National Transport Plan 2002–2011 from the four National Transport Agencies)*, Sekretariatet for Nasjonal transportplan 2002–2011, Vegdirektoratet, Oslo.

SIKA (1996), *Synspunkter på inriktningsplaneringen för väg- och järvägsinvesteringar (Review of the Socio-economic Evaluation of Road and Railway Investments in the Strategic Transport Plan 1998–2007)*, SIKA Report, p.2, SIKA, Stockholm.

SIKA (1998), Lägesanalys – *En första rapport om innriktningen av planeringen för transportinfrastrukturen 2001–2011 (Analysis of the Current Situation)*, First Report on the Policy for Planning of Transport Infrastructure 2001–2011, SIKA Report, p. 8, SIKA, Stockholm.

Statens vegvesen (1995), *Håndbok 140, Konsekvensanalyser (Handbook 140, Impact Assessment of Road Projects)*, Vegdirektoratet, Oslo.

Stenstadvold M. and Lerstang T. (1999), *Nasjonal transportplan 2002–2011, Evaluering av prosessen med fylkenes utfordringsdokument (National Transport Plan for Norway 2002–2011, Evaluation at the Planning Process at the Regional Level)*, TØI Note 1138/1999, Institute of Transport Economics, Oslo.

Trafikministeriet (1993a), *Trafik 2005 – Problemstillinger, mål og strategier (Transport 2005 – Problems, Goals and Strategies)*, Trafikministeriet, Copenhagen.

Trafikministeriet (1993b), *Trafik 2005 – Trafikpolitisk redegjørelse (Transport 2005 – Transport Policy)*, Trafikministeriet, Copenhagen.

Transportrådet (1993), *4 baggrundsnotater til Trafik 2005 (4 Background Notes for Transport 2005)*, Note No. 93–09, Transportrådet, Copenhagen.

Transportrådet (1999), *Beslutningsgrundlag for trafikinvesteringer (A Decision-making Basis for Transport Investment)*, Report No. 99–01, Transportrådet, Copenhagen.

Vägverket (1998), *Utvärdering av planeringsprocessen vid framtagendet av Nationell Plan för vägtransportsystemet 1998–2007 (Evaluation of the Planning Process for the National Road Transport System Plan 1998–2007)*, Publication, p. 79, Vägverket/Consultus, Borlänge/Stockholm.

Chapter 3

Old Myths and New Realities of Transport Infrastructure Assessment: Implications for EU Interventions in Central Europe

Deike Peters

Transport has long been considered central to urban and regional development, and to the concept of economic growth in general. The establishment of trade routes is an essential prerequisite for the expansion of market ranges. And as settlements grow, an efficient internal organisation of infrastructures is key to facilitate the efficient movements of people and goods. Distance and density remain the two most fundamental concepts in urban development theory. But of course today's cities and regions are vastly different from the places we used to live in as little as fifty years ago and consequently the relationship between transport infrastructures and economic development has been dramatically redefined over the last half century. As we have moved from Fordist to post-Fordist and post-industrial production and consumption structures, wealth creation as a whole has become an increasingly complex and de-materialised concept. Successful modern economies are less and less based on heavy production industries dependent on large-scale flow of material goods and bulk production. Instead, they rely on complex webs of inter-dependent service sector industries that are much more dependent on flexible, small batch production and just in-time delivery. Technological innovations have also vastly expanded the range of transport options, with traditional land-based modes of transportation such as road and rail now intensely competing both among themselves and with water and air transport.

Of course not everyone has drawn the same conclusions from the new realities unfolding around us. While everyone seems to agree that investments into physical rail and road infrastructures will remain crucially important even in the information age, opinions about investment priorities differ widely. This focuses on transport investment priorities in the European Union. One of the ambitions is to lay to rest the notion that a concentration of investments on large-scale links will provide economic development benefits for all of Europe and foster greater cohesion across the continent. In particular, the article evaluates European Union regional policy debates that erupted after the adoption of the Trans-European Networks (TENs) concept. The TENs are masterplans of large-scale road and rail infrastructure links that are to connect all major regional agglomerations in the European Union. The European Commission argued that the completion of the TENs was essential for making Europe globally competitive and for enhancing the development potential of Europe's peripheral regions. This first discusses general theories on the economic aspects of transport investments and then traces the recent disputes over the TENs and over European regional development policy as a whole.

Macroeconomic (non-spatial) Aspects of Transport Investments

Much confusion persists over the true growth potentials of infrastructure investments. The simplistic notion that regions will necessarily grow stronger with increased trade and interaction does not carefully enough distinguish between the economic interests of (a) existing businesses in a particular region whose production and service costs will be affected by the improvement of infrastructures; (b) consumers/employees inside the region; (c) businesses outside the region who supply the region with products and services and/or receive products, services or labour from it (or wish to do so in the future), and (d) consumers/labourers outside the region. All too often, economic growth potential is only assessed in terms of the benefits accrued by local firms, not residents. It is thus worth considering several macro-economic dynamics in more detail.

Reduction of Transport Cost

Traditionally, the reduction of travel costs has been the most important factor by which economists calculated the economic benefits of road improvements. In theory, road investments are supposed to lead to an overall lowering of transport costs, most notably through the reduction of delivery and access times. Improved transport conditions increase the market range for producers. However, in mature economies such as Western Europe and the US, this has rarely proven to be a significant factor in production patterns (SACTRA, 1999). Even the European Commission now admits that transport costs represent less than five percent of the total production cost of goods and services in Europe (CEC, 1999). Button (1993) also stresses the declining importance of transport costs and estimates that they now account for as little as two to three percent of total costs. For the U.S. context, Boarnet (1997) also points out that previous studies often overestimated the macroeconomic effects of road investments by counting in both the reduced travel costs for producers and the resulting lower consumer prices, hence in effect double-counting the benefit.

An even more important point is that, in post-industrial societies, products are increasingly high-value goods and the resulting transport cost component is relatively low. This does not mean that transport infrastructures overall are less important. But the point is that the demand for higher quality, faster infrastructure is more important than a quantitative improvement of infrastructures per se. It should also be noted that the rise of telecommunication has not necessarily replaced travel with electronic transactions but in many instances even created new demand for movement.

Even in cases where production costs are lowered due to a fall in transport costs caused by infrastructure investment, the output effect depends on whether the fall in costs is passed through to prices and, if it is, on the price elasticity of demand. Thus market elasticity and competitive conditions in the transport-using sector are critical determinants of the outcome.

Inter-Regional Trade Theory

In some cases, poor accessibility may actually benefit and protect businesses in weaker regions. In these cases, better transport connections open local markets to the increased competition of larger businesses outside the regions that produce a greater variety of products at lower prices. In other words, the regional monopoly power of existing businesses is reduced. This is a key component of the 'wider economic benefits' story. In these cases, local consumers benefit, but the region as a whole may be negatively affected by the local non-competitive businesses who struggle and begin to fail. The negative employment effects from failing local businesses are likely to outweigh consumer benefits from lower prices. Opening the region up to intra-regional competition therefore requires at least some local sectors to be strong enough to survive the competition. Given the increasing integration of European economies both within current Member States and with Central Eastern European countries, regional isolation is increasingly a non-option for European regions, but regional economists need to carefully analyse the particular strengths and weaknesses of the region in question before transport-investment-led economic development strategies can be successfully employed. Otherwise, investments might exacerbate existing divergence. As P. Martin (1999, p.12) notes: 'recent models of geographic economics show that regional integration, by reducing transaction costs between the regions, may lead to self-sustaining inequality.'

Production Function Studies: 'Is Public Expenditure Productive?'

In the late eighties, especially in the U.S., the discussion over transport infrastructure benefits shifted away from individual project analysis to studies that sought to assess the larger relationship between total infrastructure stock in a region and its economic performance. This production function approach was inspired by the larger debate spawned by Aschauer's (1989) provocative public capital hypothesis. His core argument was that public investments created benefits that spilled over to many areas distant from the actual project. These spillover benefits, it was believed, could be demonstrated by taking a more network-oriented approach, that looks at how a particular road (or rail) investment changes accessibility dynamics in other parts of the road (or rail) network.

The argument obviously has large political implications. If a general relationship between infrastructure investment and economic wealth could be plausibly demonstrated, then an expanded government role in stimulating investment into public capital was justified. The narrower neo-classical approaches focusing on cost-benefit analyses at the project level then could be said to underestimate the economic effects of public investments.

In the specific context of the U.S., Aschauer, and others after him, argued that the downturn in economic productivity in the U.S. in the early seventies could largely be explained by the overall decrease in (public) infrastructure investment since the 1960s. Consequently, they argued, only a pro-public investment approach would be able to bring growth rates in the U.S. back up to previous levels. So the supposed positive relationship between increases in public infrastructure stock and the productivity of private sector capital was largely asserted through negative deduction. Aschauer's conclusions were immediately contradicted by a number of authors, both on empirical and more ideological grounds. Interestingly, while several authors arrived at similar results as Aschauer, Ford and Poret (1991) employed a similar methodology but expanded the study to include longer data over longer time periods and for several additional OECD countries and they end up rejecting the public capital hypothesis. Other authors have attacked Aschauer on methodological grounds. Aschauer's focus on total capital stock, i.e. an exclusive focus on quantitative infrastructure measures, versus the qualitative aspects of the infrastructure in question is especially problematic. Gramlich (1994) presents an important review essay on the debates. Boarnet's (1997) more recent review article on highways and economic productivity also discusses implications for policy reform in the U.S. context. R. Martin (1999, p.143ff) discusses several mid-nineties studies that lend support to Aschauer's hypothesis for the EU context. In particular, Seitz and Licht (1995) look at the West German *Länder* during the 1971 and 1988 period, and Mas *et al* (1996) look at Spain. However, the specific emphasis in these studies is on whether public capital has cost-reducing and productivity-enhancing effects on the private sector. In other words, their interest is only in whether private firms benefit, not the resident populations. Wherever these studies find evidence for positive inter-regional spillover effects from network enhancement, they also find that these effects decline over time.

In the end, Aschauer, while providing an important ideological boost to pro-investment professionals and politicians, is not very useful for helping decision-makers with either the location or even the concrete nature of the investments. And as Boarnet (1997) noted: 'The most reasonable interpretation of the production function literature is that the U.S. infrastructure stock is well developed, such that further public capital investment will have little additional economic impact.' By limiting the discussion to an aggregate analysis of total investment stock, we learn nothing about whether, for example, road investments are more 'productive' than rail investments or whether investments are more 'productive' in core or peripheral regions. These questions, however, lie at the heart of EU regional development policy.

Transport Investment and Regional Development in the EU: Past Trends

The European Union has a legal obligation to promote balanced economic development within its territory. As the Treaty on the European Community specifies:

> In order to promote its overall harmonious development, the Community shall develop and pursue its actions leading to the strengthening of its economic and social cohesion. In particular, the Community shall aim at reducing disparities between the levels of development of the various regions and the backwardness of the least favoured regions or islands, including rural areas.[1]

The Treaty also specifies that every three years, the European Commission has to submit a report to the Parliament and the Council on 'the progress made towards achieving economic and social cohesion and on the manner in which the various means provided for in this article have contributed to it'.[2]

From the start, EU regional policy was strongly inspired by the notion that investments in transport infrastructures could help weaker regions to converge towards average levels of income in the EU. Several Commission-funded research projects assessed a strong relationship between higher transport infrastructure endowment and regional wealth as measured by GDP per capita (see especially Biehl, 1986).[3] Consequently, official EU documents, with the Commission White Paper on Growth, Competitiveness and Employment being the most prominent among them, were ostensibly justified in assuming a strong link between new infrastructure investment, growth and employment. According to this logic, an expansion in a region's infrastructure stock would result in better economic performance. And, as hinted above, stressing the macro-economic effects of infrastructure investments was also a convenient way for pro-investment politicians to avoid more extensive and more specific discussions over the particular benefits of infrastructure projects (or rather their costs and externalities). Despite the instant controversiality of the concept (and although he is usually not referred to by name), Aschauer's public capital hypothesis boosted pro-investment thinking in the EU. In particular, official EU policy soon asserted that large spillover effects and network benefits would arise from the elimination of so-called missing links and bottlenecks, particularly in border regions.

The EU Structural Funds had begun to account for an increasing share of the overall EU budget since the 1980s, and even before the advent of the TEN, transport infrastructure funding was a key element in these funds. As a Commission communication on the Common Transport Policy (CEC, 1995) put it:

> Efficient and sustainable transport systems play a key role in regional development. Structural policies and the CTP complement one another and therefore promote a more balanced and sustainable development of the Union's territory, particularly by improving accessibility and the situation of weaker regions and disadvantaged social groups.

Between 1975 and 1993, transport sector allocations through the European Regional Development and Cohesion Funds (i.e. the Structural Funds) amounted to 13.6 billion ECU in the EU. In the post-Maastricht era, transport funding increased even more substantially with the emergence of the TEN master plans. In the 1994–1999 programming period, TENs projects alone received grant support from European Regional Development Fund (EDRF) Objective 1 and Cohesion Fund grant allocations totalling 3.5 and 5 billion ECU in the 1994–1999 programming period, respectively. Another 1.34 billion were allocated through a special TEN budget line between 1995–1998. Additionally, the European Investment Bank

(EIB), the house bank of the European Union, concluded loan agreements in the amount of 14.2 billion for TEN transport projects within EU territory in the period from 1994–1997 (all figures quoted in CEC, 1999). For the period from 2000 to 2006, the Commission expects TEN budget allocation to be around 5 billion ECU.

Although public-private partnership was encouraged from the start, the TENs were always at least in part supposed to be financed by the European Union. In fact, the EU itself saw its own role as indispensable:

> Only the European Union can make the integrated transport network a reality in time to avoid the environmental and mobility crisis which faces us. (...) An integrated Trans-European transport network will bring economic and social benefits to all of Europe. It will play an important part in easing the long-term job crisis, it will be good for the environment, and it will improve the quality of people's lives (EC, The Trans-European Networks, Conclusion, 1995).

With reference to the TENs, several additional arguments were made. The European Investment Bank even used the clearly temporally limited employment effect of new construction as an argument for the TENs. According to the EIB, the TENs are 'a sound investment' for the following reasons:[4]

(1) Secure and new jobs – lower transport costs, close economic co-operation, and better development prospects for outlying regions increase Europe's prosperity and, during the construction phases of Trans-European networks, create new jobs.
(2) Greater traffic safety – modern traffic management systems and the removal of traffic bottlenecks will ensure a smooth flow of traffic and reduce the number of accidents.
(3) Less traffic congestion – less traffic congestion reduces stress, delays, costs, excessive energy consumption and pollution.
(4) Less pollution – switching traffic from roads to Trans-European rail and also waterway networks reduces pollution.
(5) Greater choice for travellers – high-speed trains cut journey times by half and offer a safe, time-saving and environmentally friendly alternative to private cars and intra-Community air travel.

> [...] These benefits are compelling. The European Union, together with the European Investment Bank and the European Investment Fund, wants to develop these Trans-European networks as swiftly as possible, for the benefit of all citizens of the European Union.

As discussed in this paper, these assumed benefits are certainly questionable to both environmentalists and even mainstream economists. As a matter of fact, the EIB's own Chief Economist's Department has just published a double issue essay collection that rethinks notions of regional development and convergence in Europe.[5] The employment effects of road and rail construction are clearly temporary and hence not able to bring permanent employment benefits to a region. Safety improvements may occur, but the early TENs visions were primarily geared at expansion, not technical upgrading of networks. As for congestion and pollution effects, the overall aim of the TENs was clearly to accommodate increasing transport volumes, which means increasing emissions overall, regardless of the fact that traffic flow improvements at bottlenecks might ease the situation at particular places. And since the supposed modal shift from road to rail is a purely rhetorical statement contradicted by the record of actual investments, it seems fair to say that the TENs in fact perpetuated road dependency in Europe, thus clearly contradicting recent EU commitments to increase the shares of rail and combined transport. In particular, the European Union's new Sustainable Development Strategy, released just in time for the Gothenburg European Summit in June 2001, included very specific goals for the transport sector. Among other things, the EU proposed to shift from road to rail, water and public transport 'so that the share of road transport in 2010 is no greater than in 1998' (CEC, 2001, p.12). There is thus rising awareness within the EU that transport has become too dependent on road travel. Another issue is the fact that increased possibilities for travel also encourage greater overall mobility, hence further increasing the overall vehicle miles travelled. Unless these additional miles travelled also result in concrete social or economic benefits that are not outweighed by additional environmental impacts, this additional kind of mobility 'for the sake of mobility' is unwelcome. In fact, the above-mentioned Sustainable Development Strategy is the first key EU document to go even one step further by making the de-coupling of transport growth from GDP growth the first headline objective for the transport sector as a whole.

Prioritising Investments at the European, National and Regional Level

From a public welfare perspective, the key consideration for an investment is the optimal use of public funds to the greatest possibly benefit of all populations in the affected regions. One therefore also has to ask the question whether new investment into transport infrastructures should be the priority strategy at all. Other investments, especially road upgrading and maintenance often promise greater economic returns (World Bank, 1996). Here, regional interest may collide with national interests. For example, local residents may consider an upgrading of the local road or rail system much more important and preferable than the construction of an additional long-distance freeway routed through their region, especially if the project cuts though valuable nature reserves. Yet, national priority projects are typically not based on the specific needs of the regions, but rather on a more general national master plan. Large-scale links often receive priority primarily not because they can be justified either through careful cost-benefit analysis or production function calculations but simply for political reasons. In the case of Germany, a large number of national priority transport projects were developed for former East Germany. And although there was indeed a large infrastructure deficit in the new *Länder*, and much investment was needed, individual projects were not necessarily responding to carefully evaluated regional needs. The demand for new infrastructure or added capacity is of course typically highest in regions with higher population densities, i.e. in core regions. Since cost benefit analyses continue to calculate the bulk of their benefits from time saving from the users, investments in these more urbanised, higher density core regions will always show greater benefits than in lower density regions. In other words, if investments were to simply follow existing demand, then peripheral regions, that are typically more rural in character, would be unlikely to receive priority investments. There is an obvious dilemma here and it demonstrates that transport policy has to be developed in consistency with spatial and regional policies. Despite efficiency arguments favouring investments in agglomerations in the core, the EU is likely to continue to be committed to developing rural and peripheral regions, and will therefore continue to finance rural transport infrastructures, which are primarily road-based. It should be clear, however, that this is first and foremost a political commitment, and not an efficiency decision. With populations and economic activity necessarily being more dispersed in peripheral regions, higher per capita investments are needed there. And given current political dynamics, with non-central regions gaining rather than losing political influence in Brussels, the EU is unlikely to discourage rural development for efficiency reasons in the near future.

Finally there is the simple and legitimate option of non-transport infrastructure investments. Especially in the mature economies of the European Union, it often makes much sense to concentrate on other aspects. For example, in order to attract the kind of highly educated labour force that the new telecommunications and computer industries require, place-factors such as cultural institutions etc. may be more important than additional access roads.

Rethinking the Theory on Transport Investments

State-of-the-art regional development theory, including most recent research funded by the EU itself, now offers a much more careful assessment of the economic benefits of transport investments than it did ten years ago. This is not only due to the overall rethinking of regional policy, but also relates to an earlier shift in emphasis from merely large-scale infrastructure investments to high-quality ones. The EU's recognition that the greatest economic benefits were likely to result from closing gaps in infrastructure connections in (centrally located) border regions rather than from a higher endowment of infrastructure stock in peripheral regions already indicated a significant shift in thinking away from a merely quantitative approach to a more qualitative one.

Additionally, recent empirical studies emerging from a variety of EU countries, particularly from those countries that have received large sums of Cohesion Fund aid for road construction, provide strong counter-evidence against the notion that large-scale transport infrastructures are a panacea for regional development. Even early-on, some European researchers warned that correlation between transport endowment and regional wealth were more likely to point to a historical relationship due to long-term agglomeration processes and did not necessarily reflect recent infrastructure investments (Brocker and Peschel, 1988). A London School of Economics study on the socio-economic impacts of transport projects financed by the Cohesion Fund found that the actual impacts varied widely, even for carefully selected infrastructure investments.

Several recent studies in particular cast serious doubt on the notion that weaker regions always benefit from transport investments (see especially Vickerman,

Spiekermann and Wegener, 1999). Even those scholars who continue to believe that transport investments are a key trigger for economic benefits increasingly admit that these benefits will not necessarily be reaped by the majority of inhabitants in the disadvantaged regions.

In its important recent Communication on Cohesion and Transport (CEC, 1999), the Commission thus still stresses the link between economic development and transport, but then goes on to admit that the link requires close examination. Even the European Commission recently acknowledged that the Italian Mezzogiorno presents a strong counterexample of a lagging region where a relatively high endowment with transport infrastructure has not as of yet resulted in a significant rise of local incomes, thereby strengthening the frequent assertion that other socio-economic factors are more important.

Meanwhile, the Nordic countries present the opposite case of highly successful regions with high average incomes and relatively low-density transport infrastructures. In fact, the Nordic countries are an interesting reminder that transport is a derived demand and that the key comparison for inter-regional or international comparisons is infrastructure endowment per capita, rather than per square mile or kilometre. Comparing the Nordic countries to the EU 15 average, it is obvious that the Nordic countries have significantly less roads and railways per square kilometre, a fact easily explained by the lower population densities (see Figure 3.1). However, both passenger and freight mileage per year are larger in the Nordic countries, owing to the simple fact that larger distances must be overcome. Interestingly, the so-called peripheral countries in the North also have higher GDPs per capita than the EU average.

In examining the difficult question whether investments into large-scale transport links bring economic benefits, both costs and benefits therefore need to be carefully assessed on a case-by-case basis. Even if an investment comes up positive in the end, this still only means that from the perspective of the project promoters, the benefits were greater than the costs. There still might be losers. And depending on the inclusiveness of the cost-benefit methodologies employed, significant losses may be externalised. Often economic benefits are reaped at the expense of greater environmental impacts, such as air pollution, which raise health care expenditures.

Interestingly, the narrowness of cost-benefit approaches has been attacked from a number of different camps, and often for opposite political reasons. Environmental externality scholars attack cost benefit analysis not for the benefits but for the costs that remain unaccounted for in traditional analyses. In the case of new toll roads, for example, environmentalists attempting to block investments will point to increasing traffic congestion and raising pollution on parallel networks due to shifted traffic onto (toll-free) roads. Meanwhile, investment boosters attempting to leverage additional highway investments that cannot be justified by narrow project analysis alone will attack cost-benefit analysis for not counting macro benefits. Rural politicians are still particularly prone to hold on to the mistaken notion that roads alleviate poverty by automatically bringing employment and growth.

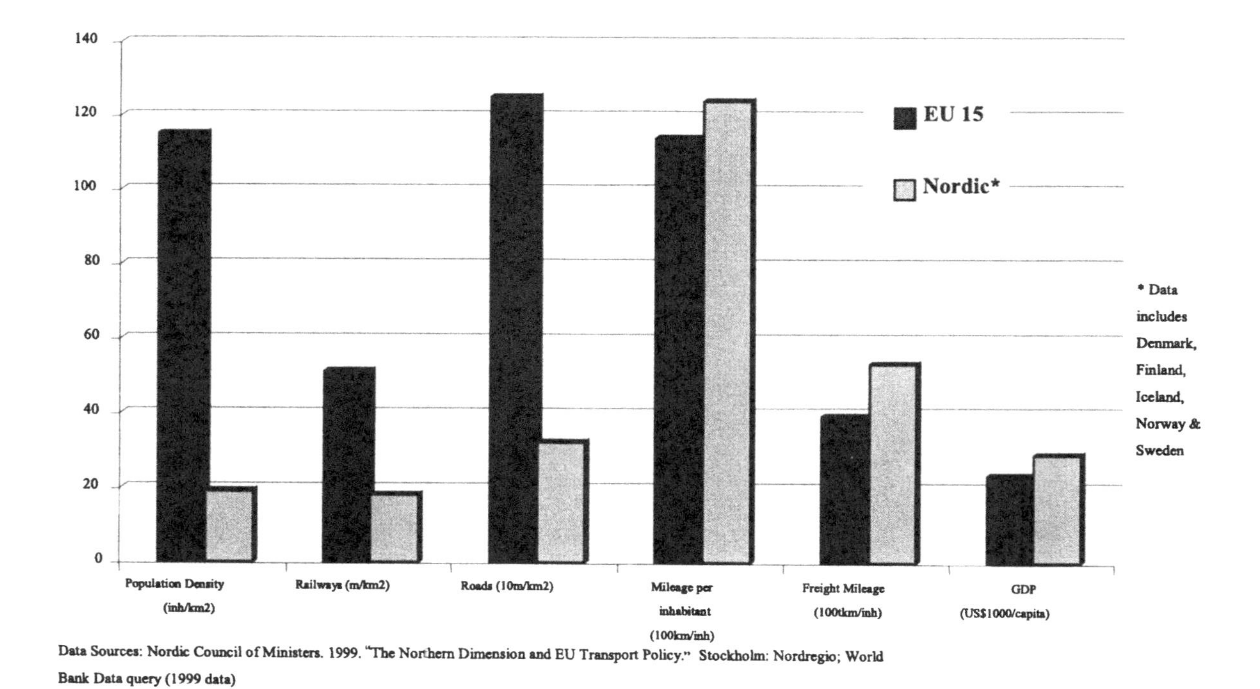

Data Sources: Nordic Council of Ministers. 1999. "The Northern Dimension and EU Transport Policy." Stockholm: Nordregio; World Bank Data query (1999 data)

Figure 3.1 The EU Versus the Nordic Countries Compared

It should also be noted that as long as the external costs of transport, particularly for freight, remain as highly externalised as in Europe (see especially Teufel, 1989), easier transport access is likely to allow manufacturers to use production inputs from ever far away places. The longer the distances travelled, the larger the environmental impact. The irrationality of using increasingly diverse inputs was underscored by Böge (1995), who demonstrated that if all product relationships were examined, i.e. if one included all the distances the different ingredients and packaging materials have to travel in order to deliver the finished product, one 'theoretical' truck of simple 150g strawberry yoghurt pots was moved over a thousand kilometres to supply her study area in Southern Germany.

Böge's analysis shatters another often-heard pro-investment argument, namely that large-scale investments will bring about a more 'rational' land use system by reorganising production and distribution systems in more efficient ways. This statement typically triggers a heated debate over what exactly an 'optimal' organisation of a land use and transportation system would be – a debate that we can scarcely afford to enter at length here. However, the following key issues should be kept in mind with respect to regional transport infrastructure investments:

First, a particular investment might ease congestion costs and hence improve travel times in the short run. However, by making travel cheaper, the investment is then also likely to encourage longer journey-making in the medium to long-term, hence inducing additional travel. And again, additional travel without concrete additional benefits is increasingly seen as undesirable. Depending on its nature, an investment might either encourage a greater clustering of activities around certain locations (agglomeration effects) or encourage a greater dispersal of economic activity (sprawl effect). The latter is generally considered to be both less efficient from a land-use perspective and responsible for more externalities. Effects can be contradictory, however. Another key analytical question is whether the investment triggers an overall increase of economic activity (win-win situation) or a just re-location of economic activity from one place to another (win-lose situation). The latter case is the essence of the so-called two-way road argument, which warns that 'improved accessibility between two countries (and similarly, between cities, areas or regions) may sometimes benefit one of them to the disbenefit of the other.' (SACTRA, 1999a, p.2). Finally, there is one last non-resolvable methodological challenge that brings us full circle and goes beyond challenging the productivity of transport investments to give recognition to the simple fact that travel itself can be both productive or unproductive, so that, depending on its nature, either an encouragement or a discouragement of movement is in order. This dilemma is most succinctly summarised in the comprehensive SACTRA report (1999a, pp.2–3), which is worth quoting at length:

> [The authors of the study] have not found it possible or helpful to define an absolute distinction between 'productive' and 'unproductive' classes of vehicles or traveller. [T]heory suggests that there are a number of important mechanisms by which (...) transport improvement could, in principle, improve economic performance. (...) In the search for empirical evidence, we find that direct statistical and case-study evidence on the size and nature of the effects of transport costs changes is limited. (...) The state of the art of this important field is poorly developed and the results do no offer convincing general evidence of the size, nature or direction of local economic impacts. (...) Our studies

underline the conclusion that generalisations about the effects of transport on the economy are subject to strong dependence on specific local circumstances and conditions.

The Paradox of European Regional Policy

The accession of several lower income countries into the EU in the 1980s made it politically necessary to institute a system of regional transfers. Resources for regional policies from the EU increased almost tenfold from 3.7 billion ECU in 1985 to 33 billion in 1999, amounting to 0.45 per cent of the EU's total GDP (P. Martin, 1999, p.11). Despite this enormous commitment, the actual economic justifications behind these substantial transfers remain contested. For one, neo-classical theorists, who assume perfect competition, would argue that policy interventions in favour of lagging regions are not justified since the process of integration itself will sufficiently accelerate convergence between regions. In other words, neo-classical theories of international trade would argue that even regions with lower level of productivity would still gain from trade based on comparative advantage. This argument is contradicted by new theories of economic geography and of endogenous growth (see especially Krugman, 1991) and by the two-way roads argument in particular. These theories emphasise the importance of economies of scale, imperfect competition and the localised nature of spillover effects. Yet ironically, new economic geography ends up challenging the validity of regional transfers in a much more troubling manner. The question is whether cohesion really is still a desirable goal even if might jeopardise European competitiveness in the long run. This forces decision-makers to clarify their objectives. Using slightly inaccessible economistic language, Philippe Martin's (1999, p.12) recent prize winning essay summarises the key dilemma:

> If economies of scale and localised spillovers explain phenomena of increased regional inequalities, this necessarily implies that efficiency gains accrue from the existence of economic agglomeration. The existence of these beneficial effects of agglomeration suggest rather that, in certain respects, Europe's economic geography is insufficiently agglomerated and specialised (for example in comparison with American geography). *It is therefore illogical to claim that the diminution of regional inequalities supposedly facilitated by regional policies will generate efficiency gains at pan-European level.* To oppose concentration and geographical specialisation is also to renounce their beneficial effects.

The key point is that a more equal distribution of economic activity across space would ultimately mean foregoing the very benefits that urban and regional agglomerations provide. In other words, although hugely problematic from an equity perspective, the argument is that Europe as a whole may be better off economically if it kept concentrating infrastructure investments in the Blue Banana core of Europe, with certain peripheral exceptions. Martin is thus making an efficiency argument that the EU, for political reasons, could never heed.[6] Ironically, the catchy Blue[7] Banana image of an economic backbone corridor in Europe reaching from London across the channel through the Benelux countries, Northern France, Switzerland and Southern Germany to Northern Italy was first developed by

French spatial analysts in the late 1980s precisely in order to point out the need to develop an alternative, more polycentric structure (also see Figure 3.2). This need certainly has also been the focus of the European Spatial Development Perspective (ESDP) document that features the ambitious subtitle 'Towards Balanced and Sustainable Development of the Territory of the EU.' The ESDP was developed following the adoption of the so-called Leipzig Principles by European ministers in 1994. The principles call for: (a) a balanced and polycentric city system and a new urban-rural partnership (b) parity of access to infrastructure and knowledge, and (c) sustainable development, prudent management and protection of nature and cultural heritage (listed in Faludi, 2000, p.8). The two different visions of Europe are nicely illustrated in Kunzmann's sketch of the European Blue Banana versus the European Grape (Figure 3.3).

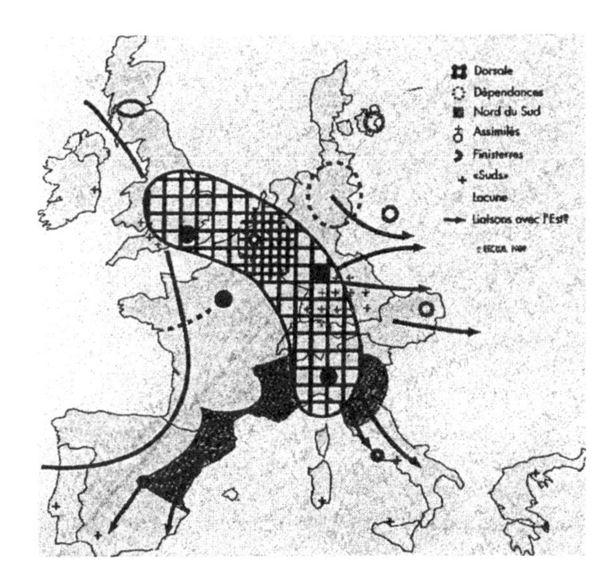

Figure 3.2 The Blue Banana Concept

Source: Reclus 1989, Reproduced in NSPA 2000

To suggest that the very idea of EU Cohesion and/or Spatial Development policy, namely to achieve a more balanced distribution of infrastructure across space, is contradictory to what the EU's own house bank and main lending institution (i.e. the EIB) now considers state of the art economic rationality is indeed a deeply troubling thought, especially since the EIB itself has been the main funding institution for those very TEN networks that supposedly will be developed 'for the benefit of all citizens of the European Union'.[8] The contrasting visions also explain the recurring dilemma of all decision-making surrounding large-scale infrastructures: by

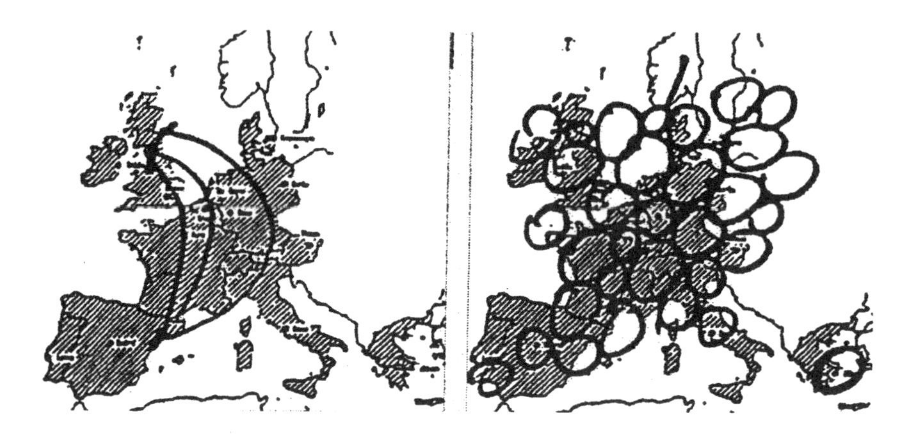

Figure 3.3 Competing Fruits: The European Banana Versus the European Grape

Source: Kunzmann 1996, Reproduced in: NSPA0

definition, all 'backbone networks' – be they located along the Blue Banana or along any of the ten chosen Helsinki Pan-European corridors – privilege connections between large cities and bypass agglomerations of lesser importance. (And the bypass effect is certainly much greater in the case of high-speed rail than for roads.) This means losers are scattered along the way, and, according to the SACTRA experts, even possibly also at one of the two ends. This latter point becomes particularly important when we consider the EU's ambitions to extend the TENs into Central and Eastern Europe.

Modal Bias in the TENs: Road Continues to Win Over Rail

Another myth about the TENs was the idea that they would primarily be focusing on high-speed rail links. By favouring rail over road, transport systems would supposedly become more sustainable. Leaving aside for the moment the argument that, depending on how the accounting is done and which costs and benefits are included, new high-speed trains are not necessarily more energy efficient or environmentally friendly than road traffic, the Commission's rhetoric of actively pursuing a modal shift from road to rail is simply not backed up by reality. In fact, data analysis on transport investments shows the Commission until now has failed to meet its own recommendation to promote environmentally friendlier modes.

It is true that 9 out of the EU's 14 TEN priority projects were high-speed rail projects (see Figure 3.8). Also, over 60 per cent (827 out of 1344 million ECU) from the TEN special budget line went towards (high speed) rail. However, Figure 3.4

also indicates that the vast majority of transport spending under the more sizeable Cohesion and ERDF Funds went towards roads, tipping the overall balance about two-thirds in favour of roads and highways. From 1994–1999, over 70 percent of the 13.7 available ERDF Objective 1 funding went to roads.[9] The imbalance was equally pronounced in the case of the Cohesion Fund. From 1993 to 1999, TEN-related priority transport investments to the four poorest member states (Greece, Ireland, Portugal and Spain) accounted for over 5 billion Ecu, of which 69 per cent went to roads (see Figure 3.5).

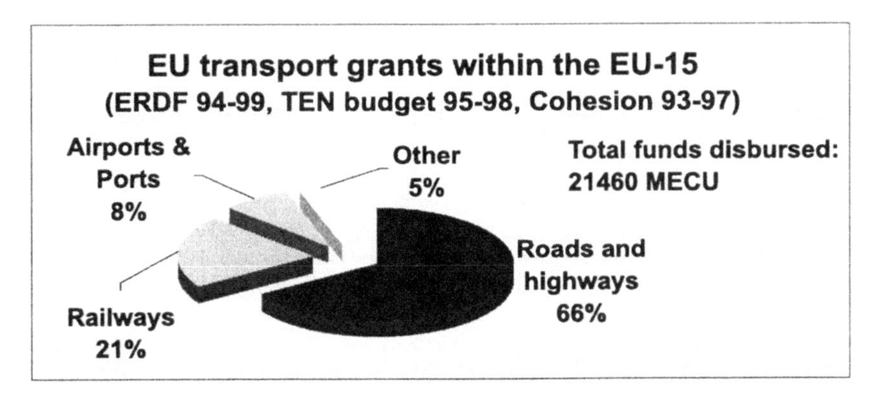

Figure 3.4 The Modal Bias in EU Transport Funding

Data Source: CEC, 1999

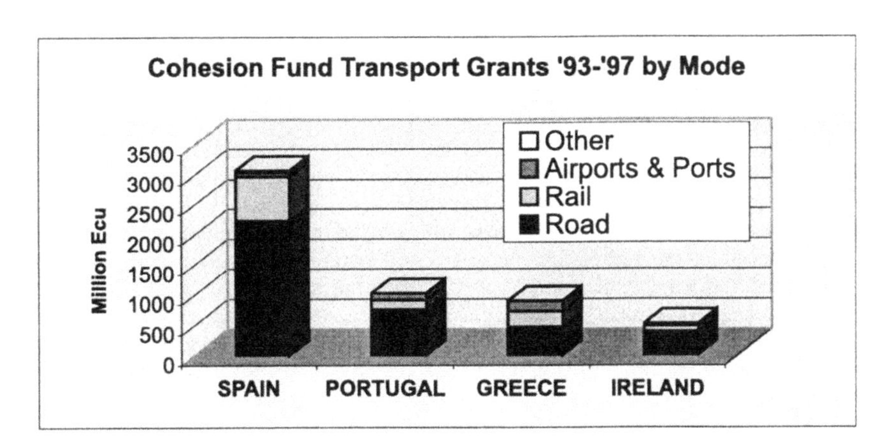

Figure 3.5 The Modal Bias in EU Cohesion Fund Grants

Data Source: CEC, 1999

The Empty Rhetoric of Cohesion and Sustainability

EU development goals remain deeply conflicted and contested. EU transport sector investments continue to have to satisfy different aims related to economic growth, regional cohesion and sustainability. Environmental concerns often take a back seat to mainstream economic development interests, and the politically most influential core regions continue to be able to attract a significant amount of infrastructure funding at the expense of less economically advanced peripheral areas. The TENs priority projects fundamentally violate cohesion goals and sustainable development by concentrating investments in already privileged areas. And despite a rhetorical favouring of rail, the majority of investments still went towards road projects. One of the key reasons why lending and grant making remained so heavily biased towards road, was the failing commitment of the recipient governments to more actively pursue a modal shift towards rail. And in the cases of the chosen fourteen TEN priority projects, which were determined at the 1995 Essen Council meeting of the EU, most of the high-profile road and rail projects were in fact long-standing pet industry projects that had been heavily promoted by the industrial lobby for some time (also see below).

Finally and perhaps most importantly, the diminished faith in transport investments as triggers of economic growth in lagging regions must be seen in the larger context of the recent re-evaluation of the possibilities of regional policy as a whole. In Europe, empirical evidence in the 1990s has been used to support both supporters and dissenters of the view that regional policies can help poorer regions catch up with wealthier ones. Many commentators argue that the economic forces leading to an increasing divergence between regions are simply too strong for regional policies to counteract them. According to this view, infrastructure investments in poorer regions appear as pure income transfers that are unlikely to seriously narrow the productivity gap between poorer and richer regions. In fact, improved transport connections may even accelerate out-migration in poor regions and thus widen the gap rather than narrow it. Alternatively, one might argue that these types of transfers negatively affect overall growth. As Philippe Maystadt (2000, p.4), the president of the EIB, recently noted: 'Indeed, [regional spending] may lead to lower overall prosperity if it drains resources from those wealthy and innovative regions that are the main engines of economic growth. If this is the case, we face a trade-off between equality and growth.'

While this may seem like news to neo-classical economists, it is a truth that political economists have long emphasised. And an increasing number of scholars concerned about the increasing environmental burden that our transport systems impose upon us would throw in ecology for good measure, arguing that it is really a three-way trade off. In sum, we are once again faced with the fact that the goal of sustainable development, regional or otherwise, is always struggling to balance three at times incompatible dimensions: economic growth, equity and the environment.

EU Realities: Lobbying Power Explains the Rapid Adoption of the TEN Concept

As above discussion shows, the TENs concept was obviously not backed up by unequivocal economic reasoning. So how could the TENs so quickly advance from a mere paper tiger to a multi-billion Euro investment programme? The best explanation is that decisions were highly political. The rapid adoption of the TENs concept in the early 1990s is primarily due to their timeliness and the ability of the concept to respond to both urgent industry and high-level political needs.

The pro-investment infrastructure and construction lobby and the EU joined forces on the idea of the TENs. In some ways, it may be even more accurate to say that most official European Union transport infrastructure investment proposals, and the priority Trans-European Network projects in particular, originated as industry lobby proposals that were only later transformed into EU policy. There are several key reports prepared for the European Roundtable of Industrialists (ERT) which resemble future Commission proposals in startling ways. The ERT's 1984 report called 'Missing Links' outlining three specific proposals for a 'Channel link between England and France,' a 'Scanlink' plan for road and rail connections in Scandinavia, and a 'Trans-European network of high-speed trains' is certainly the boldest and most forceful document, urging billions of dollar of investments.[10]

The similarity of even this early proposal with the list of TEN priority projects adopted ten years later at the Essen Summit in 1994 is striking (Figures 3.6 and 3.8). Not only did the 1994 EU list end up including the Channel tunnel and the Øresund road-rail bridge as individual priority projects, but both were also integrated into a network of high-speed rail links that picked up most of the connections originally proposed by ERT ten years earlier. Interestingly, ERT soon created the myth that 'not only are such projects desirable in terms of their economic and social impact, but they are affordable, and can be profitable, environmentally acceptable and financeable without heavy extra commitments to public spending' (p.1). In fact, ERT claimed, some of the link projects:

> could be financed in very large measure by the private sector. If governments were prepared to set the right investment climate in terms of fiscal incentives and operating licences, the money could be raised. Although Europe has become culturally programmed to see large transport infrastructure projects as the preserve of government, there is no reason why this should be so.

Corporate watchdog organisations such as ASEED Europe acknowledge the immense success of the ERT in influencing future EU transport policy. As Doherty and Hoedeman (1994, p.137) note:

> Through its intensive lobbying of European transport ministers, and also the support of French Prime Minister Laurent Fabius, the ERT was astonishingly successful in introducing European power brokers to its vision of a future infrastructure. In 1985, Volvo's Pehr Gyllenhammer could report to ERT members that the Italian government, 'on behalf of all the ministers of transportation within the community, is referring to the Missing Links as a master plan for European infrastructure'.

Richardson (1997, p.337) also provides a strong case that decisions were influenced by the privileged access that industry decision-makers had to key Commission working groups:

> Proposals for the Trans-European Road Network (...) were developed by the Motorway Working Group (MGW) of the Commission's Transport Infrastructure Committee [which included] a number of private sector interests including the European Round Table (ERT), the Association des Constructeurs Europeens d'Automobile, and the International Road Transport Union. The Committee was overwhelmingly dominated by transport and infrastructure interests, with a notable absence of environmental interests. (...) It appears that the debate within this key decision-making arena was largely political. The institutional power of the infrastructure lobby demonstrated here was strengthened by ready access to top-ranking EC and member-state politicians.

ERT lobbying for the missing links was heavily stepped up before the passing of the Maastricht Treaty, with three more ERT publications further underlining previous calls for Trans-European networks. Already two years after the ERT proposal, the European Conference of Ministers of Transport published a curious sketch-like map of Europe that showed a list of 'missing links' in European road and rail infrastructure (Figure 3.7).

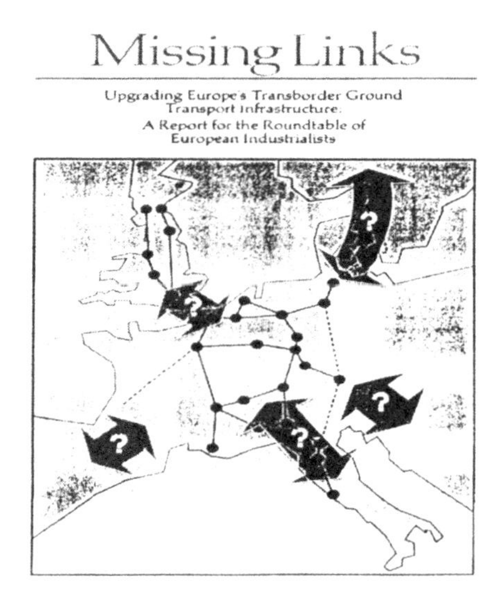

Figure 3.6 Europe's Missing Transport Infrastructure Links According to the ERT (1984)

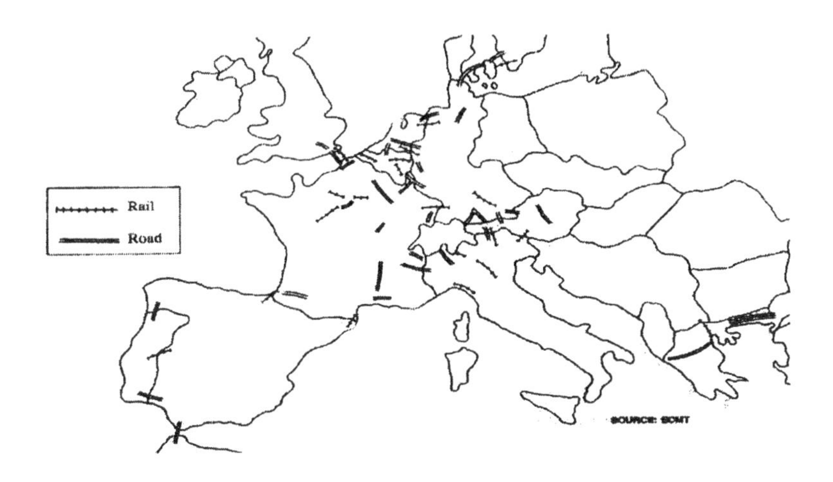

Figure 3.7 Europe's Missing Links According to the ECMT (1986)

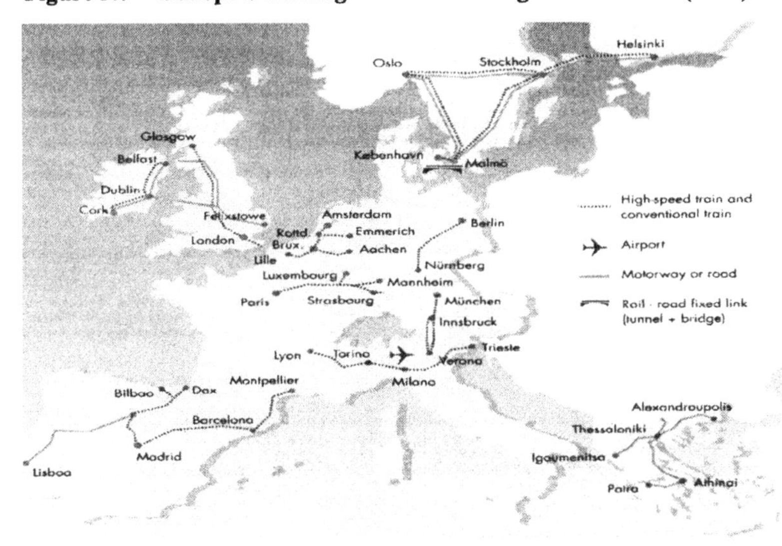

Figure 3.8 The 14 TEN Priority Projects Adopted in Essen in 1994

Source: EIB, 1998

Yet while there are serious objections to the TEN on both environmental and social, and even on economic grounds, several important arguments worked in favour of the TENs from the point of view of the Commission. Besides the strong pressures from the industry in the form of the Round Table of European Industrialists, there was internal pressure in the form of what Turro (1999, p.102) calls the 'bicycle theory'. According to this theory, the European Union is constantly forced to 'keep pedalling to avoid falling off' i.e. it has to push and explore new areas of co-operation and increase political and economic co-operation in order to keep the union alive. And in the particular context of the early 1990s, embarking on the TENs served several strategic objectives for the Commission: First, it provided a new 'safe' field of co-operation at a time where it was too early to seriously push common defence and foreign policy or monetary unification. Second and more importantly, Trans-European infrastructures could be used to justify an important increase in the Community budget and a reduction of the excessive share of the overall budget that was spent on the Common Agricultural Policy (CAP). Since the implementation of the projects would remain a national responsibility, no substantial expansion of the EU bureaucracy was needed. Turro (1999, p.103) also points out that:

> under these conditions, the transfer of investment from the national to the Community budget could be of interest to Finance Ministers needing to improve the appearance of their public debt and national deficit figures to comply with the EMU conditions. [T]he TENs concept had the rare virtue of combining national and common interests and to be timely and mostly non-controversial. (…) This explains its quick progress in relation to the normal pace of European policy-building.

While common interests and timeliness provide practical explanations for the rapid increase in TEN financing inside the EU, they of course do not provide a justification for the lack of strategic and environmental assessment that accompanied the TEN planning process.

Richardson (1997) nicely documents the institutional battle between the Council, which was largely supportive of the Commission's TEN proposal, and the European Parliament, which fought to integrate a number of key environmental provisions into the proposal, most notably a requirement for doing a Strategic Environmental Assessment (SEA) on the TENs. Although much progress has been made on developing SEA methodologies in general, the actual issue of doing a full-scale SEA on the TENs remains unresolved to date.

Expanding the TENs to Central Eastern Europe: A Whole Different can of Worms

When it comes to Central Eastern Europe (CEE) and enlargement, neither bicycle-theory nor lobby-power necessarily fully explain the form and function of EU interventions related to Trans-European Network expansion outside current EU territory. Conflicts of interests surrounding EU-CEE links are even more complex than in the case of internal EU connections between member states, and prospects for environmentally sustainable, transparent infrastructure decision-making are bleaker in the case of the candidate countries.

European Transport Ministers from both EU and Central Eastern European countries met for the first post-transition, Pan-European Ministers of Transport Conference in Prague in 1993. The first set of nine international priority corridors was subsequently adopted at the Pan-European Transport Conference in Crete in 1995.[11] One additional corridor was added during the next Pan-European conference in Helsinki in June 1997, completing what is now commonly referred to as the Helsinki Corridors.

Much like the TEN priority projects adopted in Essen, the Helsinki Corridors do not necessarily have their basis in detailed traffic counts and forecasts but were rather determined at a high political level. Not all corridors are of equal importance to Trans-European transport flows, since they largely had to be determined by what would be politically acceptable to all stakeholders. It was politically necessary to make sure that each capital city was included in at least one of the corridors. So although they are given equal weight on the resulting maps, the section of corridor II linking Berlin and Warsaw is obviously of much greater importance to Pan-European trade and traffic flows than the stretch of corridor VII linking Durres to Tirana, for example.[12]

In order to put the highly politicised process of corridor 'prioritisation' on an ostensibly more scientific and more objective footing, the European Commission, in co-operation with the candidate states, and with financial assistance from the EU

Phare Program and the City of Vienna, launched the so-called 'Transport Infrastructure Needs Assessment (TINA)' process. Even the European Commission itself readily admitted that the Helsinki Corridors largely reflected EU strategic interests, and that a larger needs assessment for the candidate countries was necessary. The first official mention of a Transport Infrastructure Needs Assessment (TINA) for the CEE candidate countries process was made at the first structured dialogue between the European Union Transport Council (of Ministers) and the relevant transport ministers of the candidate countries in 1995. At the time, the main goal was an identification of the transport-sector investments required to bring transport infrastructures in CEE countries up to EU standards and to ensure adequate linkage between CEE and EU infrastructures.

It is important to note that the TINA infrastructures are divided into a first-order backbone network that consists of an amended version of the Helsinki corridors, and a denser, second-order network that includes infrastructure of national importance. Since it is the most comprehensive and most high-profile exercise in Pan-European infrastructure assessment, TINA was always expected to form the basis for subsequent pre-accession funds in the transport sector, which explains why national governments had a strong interest in making sure all infrastructures of national importance were reflected in these maps.[13] Figures 3.9 and 3.10 show how original proposals for the Pan-European corridors further reinforced the already over-centralised Hungarian (rail) network, while subsequent inclusions of additional corridors proposed by TINA participants ensured a slightly more balanced approach to the Hungarian rail system.

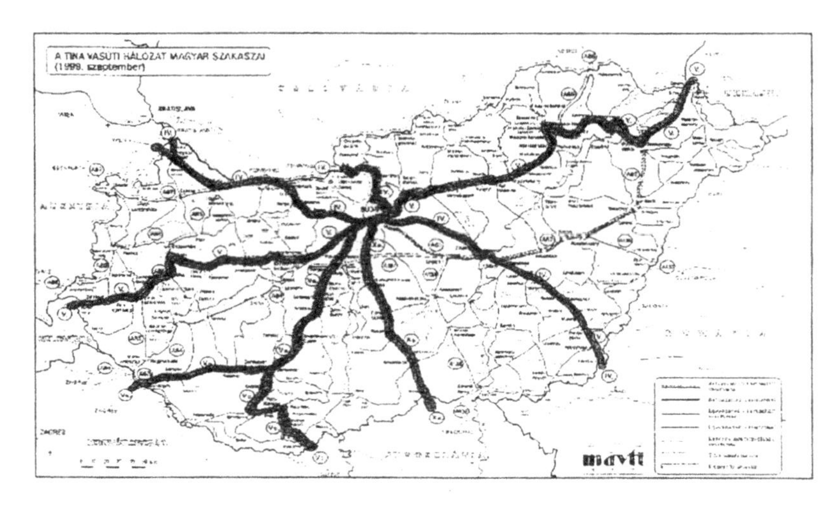

Figure 3.9 The Magistral Network of the Pan-European Corridors on the Hungarian Railway Network (Without the Additional Corridors Suggested by the Member States')

Source: Fleischer, 2000, p.25

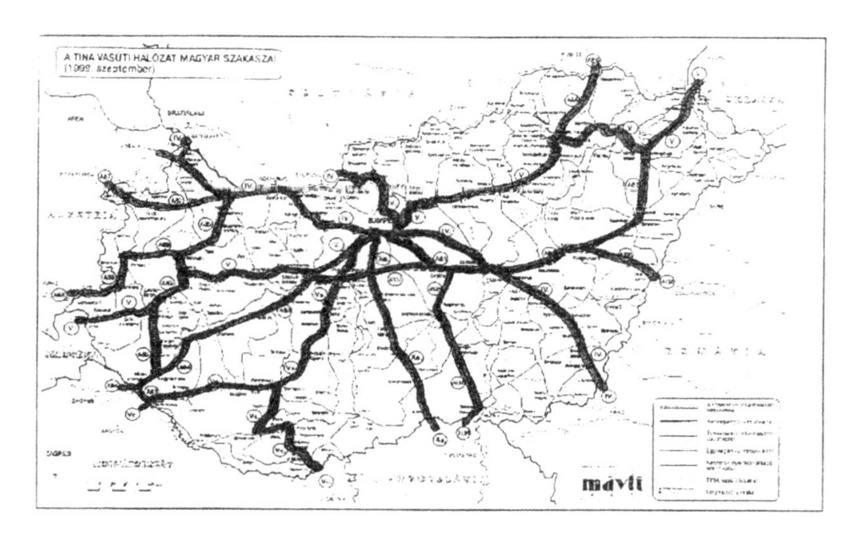

Figure 3.10 The TINA Corridors on the Hungarian Railway Network (Including the Additional Corridors Suggested by the Member States')

Source: Fleischer, 2000, p.26

The status of TINA as the guiding tool for EU investment in CEE transport infrastructure remains problematic (and actually somewhat disputed even within the Commission itself). It is also important to note that at present, in contrast to the Crete and Helsinki corridors, the TINA network has only been adopted by a senior officials group and not by any higher, i.e. ministerial level, and thus has no formal legal base within the EU.[14] The total needs listed under TINA are enormous, and meeting the investment schedule under the 2010 time horizon would involve spending more than 1.5 per cent of the country' GDP in some of the candidate countries.[15]

An SEA on TINA? The Challenge of Strategically Assessing a 'Needs Assessment'

Due to the politicised nature of the process, TINA was never really an objective needs assessment. Also, the TINA corridor maps ended up becoming a key outcome in addition to the investment charts. It should also be noted that with a time horizon of over 15 years, estimations on expected traffic flows were necessarily highly speculative.

There were several voices that called for a timely Strategic Environmental Assessment (SEA) on the TEN extensions, especially after chances to do one for the TENs were missed. In particular, the ECMT/OECD workshop on SEA in the Transport Sector called for the priority application of SEA for transport infrastructure development plans laid down in TINA recommendations (ECMT/OECD, 1999, conclusions).

The Commission subsequently commissioned an 'Environmental Scoping' on the Multi-Modal Transport Corridor VI from Warsaw to Budapest, organised by the Regional Environmental Centre (REC) in Szentendre, Hungary. It was a partial if limited response to the ECMT's somewhat ambiguously formulated call. It can hardly be regarded as an acceptable shorthand version of a strategic assessment of TINA, however.

As was to be expected, the most problematic aspect of the REC study was to decide the appropriate scope. While the original aim was to assess the environmental impacts all the transport policies in the entire CEE region, i.e. do a full 'SEA of the TINA network', this ambitious aim was quickly scaled down. As the Italian SEA expert Olivia Bina (1999) already noted during the Warsaw SEA conference: '[it is] not feasible (...) to attempt to complete a full assessment of the impacts of the network at this large scale of analysis.'

The compromise solution that was found was to focus on one multi-modal corridor instead. It was believed that 'a well chosen corridor would provide, if not representative, at least indicative evidence of the potential environmental problems faced by the rest of the region'. The general objectives of the study were defined as follows (REC, 2000, p.2):

- to identify potential impacts of the proposed development of transport infrastructure in CEE on environmental commitments of CEE countries,

- to suggest suitable tools for integration of environment and transport (with specific focus on potential application of SEA for international transport planning in CEE),
- to suggest practical provisions for effective co-ordination of environmental and transport planning on CEE regional level [sic].

However, the report also clearly acknowledged the impossibility of doing any real ex-ante evaluation, since the corridor is already a designated Pan-European corridor and as such part of the priority investment list of both national and international planning documents. The scooping therefore was to be carried out as an *ad interim* assessment, focusing on existing environmental and transport policy contexts.

The final synthesis report of the exercise begins by noting that 'the EU Accession Process is the main driver of change in Central and Eastern Europe' (REC, 2000, p.1). The EU Accession process has indeed already fundamentally transformed the environmental and transportation policies in the CEE candidate countries.[16] Several conclusions that emerged from the REC exercise are revealing. For example, the report notes that urban air pollution and loss of habitat are two major impacts directly related to rising traffic volumes, which is in turn related to higher incomes and accelerated trade between East and West.

Much more importantly, the synthesis report authors come to the clear conclusion that *'the Community support for the TENs [extensions in CEE] is drawing the limited national resources primarily towards construction of long distance infrastructure, primarily highways'* (REC, 2000, p.9, my emphasis).

REC researchers found that now that the backbone network of the Trans-European network extensions has been determined in the form of the Helsinki corridors, CEE in-country representatives frequently (and often purposely) misinterpret the TEN-extensions and the entire TINA exercise as an agreed and consultative international agreement that now needs to be implemented at the national level. The argument is often made in conjunction with the myth that the TINA backbone network is something that will largely be funded by the EU (i.e. ISPA) and the IFIs. Consequently anyone who now opposes the content and priority networks contained in TINA appears as someone who is jeopardising international funding and assistance for their countries (Dusik, 2000).[17] And since, so the interpretation frequently goes, EU and IFI co-financing is mostly available for backbone infrastructures, national priorities should consequently be redirected in the same direction, regardless of actual independently assessed needs. The reports then also support the corollary finding that TINA effectively overrode national transport policies in the sense that national-level priorities were changed according to what the TINA final report stated (Fleischer, 2000).

Finally, the reports also pointed to the fact that since ISPA is modelled on the Cohesion Fund, EU transport sector grant assistance currently cannot be given for much needed urban public transit improvements, thereby further exacerbating detrimental disinvestments trends.

The Challenge of Sustainable Infrastructure Lending in Central Eastern Europe

Current developments in Central Eastern Europe regarding the likely over-prioritisation of large-scale EU-CEE road connections to the detriment of other key national infrastructure needs may be in part a tragedy of unintended consequences. This does not make it any less problematic.

On a positive note, even if the true reality of transport corridor assessment in the 'New Europe' is likely to continue to be dominated by political opportunism and geo-strategic thinking, significant strides in developing better environmental, social and financial assessment methodologies for large-scale infrastructure investments have been made. Unfortunately, EAs, SEAs and EIAs are not always carefully and extensively enough applied, especially when it comes to EU-funded transport projects in CEE. Also, project officers at the International Financial Institutions, and at the EIB in particular, are under great pressure to perform environmental, economic and financial appraisals in extremely time-constrained situations, and officers candidly admit that even if more sophisticated and elaborate appraisal methods were available to them, most of them would likely not be applied on a regular basis unless the assessments could be completed in the necessary time frame available during a regular project cycle. Project officers are sometimes responsible for as many as 25 projects at the same time. Overall, the lending practices of the European Investment Bank have been strongly critiqued by environmentalists and civil society organisations, and by independent transport experts. EU parliamentarians, EU member state governments and international organisations are also increasingly concerned about the lack of public accountability at the EIB and are beginning to take action.[18] Considering the tremendously important role the EIB is taking on in financing TENs and TEN extensions in CEE, it is imperative that the EIB's environmental and social assessment methodologies and their public information disclosure practices become at least comparable to those of the World Bank.[19] In the end, newly developed state-of-the-art assessment methodologies are useless unless they are also rigorously applied in practice.

Notes

1 Title XVII (ex Title XIV) 'Economic and Social Cohesion', Article 158 (ex Article 130s) of the Treaty establishing the European Community, consolidated version incorporating the changes made in the Treaty of Amsterdam, published in the Official Journal C 340, 10.11.1997, pp.173–308 and online under http://www.europa.eu.int/eur-lex/en/treaties/.

2 ibid. Article 159. The report also goes to the Economic and Social Committee and the Committee and the Regions.

3 Note that the Nordic countries, which have below-average urbanisation rates but above-average national incomes, were not part of the EU in the 1980s. The case of the Nordic countries also makes clear how important it is to differentiate between infrastructure endowment per square kilometre, which would be comparatively low in the Nordic countries, and infrastructure endowment per person, which would be comparatively high in rural areas.

4 Source: EIB Website (www.eib.org), data retrieved September 2000.

5 See the essays in *EIB Papers*, 2000, volume 5, number 1 ('Regional Development in Europe: An Assessment of policy strategies') and number 2 ('Regional convergence in Europe: theory and empirical evidence').
6 It should be noted, however, that current rethinking of regional policy as a whole does not necessarily challenge the sense of using road and rail investments as tools for economic development, since neither market-led economic integration nor diversification can be achieved without first physically linking the infrastructures of countries or regions. What it does, challenge, however, is previous simplistic assertions (made both by the Commission and others) that improved road and rail connection will automatically reduce regional disparities.
7 What makes the banana blue? I always assumed that this was the colour used in the original maps produced by Brunet *et al* in their 1989 DATAR study, but I have been unable to verify this first hand.
8 Much important *grant* funding for TEN networks particularly in the four lower income EU member countries has come from the EU Cohesion Fund. However, a much larger part of the TENs was financed through favourable *loans* from the EIB, as well as through public-private partnerships and national government funds.
9 Objective 1 funding is designated for the least developed regions of the EU, i.e. regions where GDP per capita is below 75 percent of the EU average.
10 The full title is *Missing Links – Upgrading Europe's Trans-Border Ground Transport Infrastructure: A Report for the Roundtable of European Industrialists*. The glossy report (black and white copies are available at no charge from the Roundtable's Paris or London offices) contains the following stunning note in an appendix: 'The working group aimed to provide a concise and readable report. For this reason, data sources and references are conspicuously absent from the text, charts and tables. The group's report is compiled from the following written [internally commissioned] submissions and reports: [List reports] The 'pedigree' of the facts and figures quoted in the *Missing Links* report is finely detailed in the written submissions and reports listed above. Questions on the pedigree of facts and figures [--note: not orders for the pedigree itself!--] should be addressed to Michael Hinks-Edwards at the Roundtable Secretariat Paris Office (see address below).' In other words: the bold ERT proposal was based on data that is neither independently accessible nor verifiable.
11 The Dutch Ministry had been the first one to suggest that representatives focus on developing a map of transport priority corridors connecting the European Union to its eastern and southern neighbours. Thus the corridor concept was born, partly for want of any other equally appealing proposal.
12 In fact, this latter corridor in particular was added for purely political reasons.
13 However, added corridors normally did not become part of the backbone, but rather of the general TINA network. The distinction between the two (as well as the prioritisation of ISPA funding for the backbone networks rather than TINA infrastructures in general) remains a point of contention and confusion both in the EU and CEE, especially since the corridor concept as a whole is being abandoned inside the EU.
14 DG Transport and Energy representatives plan to change this status by attaching the TINA report as an annex to the planned revision of the TEN guidelines, which are formal EU law. Since these TEN guidelines are part of the so-called transport Acquis, i.e. the transport-specific sections of the Acquis Communautaire, which form the basis of the accession treaties, they thereby become binding legislation for the accession countries as well.
15 Senior Commission representatives consider this a very realistic and doable investment plan, and pride themselves in having lowered overly elevated infrastructure building cost estimates on the part of the candidate countries. In the case of Poland, Commission-hired

consultants supposedly found cost estimates for a variety of road infrastructure schemes to be lower than Polish supplied data by a factor of 2.

16 Most CEE countries have by now adopted National Environmental Policies, National Environmental Action Programs, and National Transport Policies that take into account EU requirements and regulations, as well as strategic objectives.

17 Another key issue from a procedural perspective is that TINA was still largely a 'closed room assessment' in the sense that the process only involved the transport ministries, and excluded decision-makers from the Ministries of Finance, Environment or Regional Development as well as civil society stakeholders.

18 The Central European NGO-Network Bankwatch prepared a highly critical study on the EIB for the German Heinrich Boell Foundation (see CEE Bankwatch Network 1999). On November 9–10, 2000, Green EU Parliamentarians organised a conference with the lengthy title 'Invisible Hands Shaping Europe – A closer look at the European Investment Bank's policies and projects from an environmental, access to information and public participation perspective.' Finally, the recent international workshop 'FIST – Financing Sustainable Transport Infrastructure and Technology focussing on CEEC and NIS' organised by the Austrian Ministries of Transport and Environment, the UN and the OECD's Central European Initiative produced an officially endorsed input paper for the 9th UN Commission on Sustainable Development meeting in New York in April 2001 that contains a series of specific recommendations for improving the social and environmental accountability for transport infrastructure lending at the international financial institutions (IFIs).

19 Hook, Peters *et al* (1999) prepared an extensive comparative assessment of transport lending practices at the World Bank, EBRD and the EIB (written under the auspices of the German Federal Environmental Agency). Unlike the World Bank or the European Bank for Reconstruction and Development (EBRD), the EIB has not developed any internal transport-sector policies to guide its lending activities. The EIB has repeatedly claimed that such internal lending guidelines are not necessary since the EIB is supposedly guided by EU laws and policies. However, EIB staff does not feel bound by EU policies or practices when lending outside current EU territory. Also, while the World Bank makes all of its EIAs; sector and policy papers; staff appraisal reports; project summaries and country strategies available to the public, the EBRD only makes EIAs, sector papers, project summaries and press releases available. All the EIB provides to the public are press releases. Finally, it should be noted that even the World Bank, who, responding to NGO pressure, has extensively reformed its environmental, social and public information policies in the last decade, still uses financial and economic assessment criteria that bias lending in favour of road over rail (see especially Hook, 1996).

References

Amin, A. and Tomaney J. (eds) (1995), *Behind the Myth of the European Union – Prospects for Cohesion*, Routledge, New York.

Aschauer, D.A. and Federal Reserve Bank of Chicago (1988), *Is public expenditure productive?*, Federal Reserve Bank of Chicago, Chicago, IL.

Biehl, D.E. (1986), *The Contribution of Infrastructure to Regional Development. Final Report of the Infrastructure Studies Group to the Commission of the European Communities*, Office for Official Publications of the European Communities, Luxembourg.

Bina, O. (1999), *SEA to Date, Recent Advances and Current Priorities for Development*, OECD/ECMT Conference on SEA for Transport, Warsaw, 14–15 October.

Boarnet, M. (1997), 'Highways and Economic Productivity: Interpreting recent evidence', *Journal of Planning Literature*, Vol. 11, No. 4, pp.476–486.

Böge, S. (1995), 'The Well Traveled Yogurt Pot: Lessons for new freight transport policies and regional production', *World Transport Policies and Practice* ,Vol. 1 No. 1, pp.7–11.

Brocker, J. and Peschel, K. (1988), 'Trade' in Molle, W. and Cappelin, R., *Regional Impact of Community Policies in Europe*, Avebury, Aldershot, pp.127–151.

Button, K.J. (1993), *Transport Economics*, Elgar, Aldershot, Hants, England, and Brookfield, Vt.

CEE Bankwatch Network (1999), *The European Investment Bank: Accountable Only to the Market?*, Brussels, Heinrich Böll Foundation.

Commission of the European Communities (CEC, 2001), *A Sustainable Europe for Better World, A European Union Strategy for Sustainable Development*, Brussels, European Commission.

Commission of the European Communities and European Commission (1999), *Cohesion and Transport*, Brussels.

De la Fuente, A. (2000), *Convergence across Countries and Regions, Theory and Empirics*, EIB Papers/Cahiers BEI, Vol. 5, No. 2, pp.25–36.

Doherty, A. and Hoedeman, O. (1994), 'Misshaping Europe, The European Round Table of Industrialists, *The Ecologist*, Vol. 24, No. 4, (July/August) pp.135–141.

Dusik, J. (2000), *Interview at the REC Centre in Szentendre*, August 2nd, European Commission (1995), The Trans-European Networks.

European Round Table of Industrialists (1984), *Missing Links – Upgrading Europe's Transborder Ground Transport Infrastructure*, Brussels.

European Round Table of Industrialists (1989), *Need for Renewing Transport Infrastructure in Europe*, Brussels, European Round Table of Industrialists.

European Round Table of Industrialists (1991), *Missing Networks, a European Challenge – Proposals for the Renewal of Europe's Infrastructure*, Brussels, European Round Table of Industrialists.

European Round Table of Industrialists (1992), *Growing together, one Infrastructure for Europe*, Brussels, European Round Table of Industrialists.

European Union (1997), *Treaty establishing the European Community, Consolidated Version incorporating the changes made in the Treaty of Amsterdam, Official Journal*, C 340, pp.173–308.

Faludi, A. (2000), *Fools dare where Angels fear to tread?* Paper presented at the Annual ACSP Conference, Atlanta, USA, November 1–5.

Fleischer, T. (2000), *Initial Scoping Phase of a Strategic Environmental Assessment (SEA) of the Multi-Modal Transport Corridor Warsaw-Budapest*, Final Draft of In-country Scoping Report, Hungary, Szentendre, REC.

Ford, R. and Poret, P. (1991), 'Infrastructure and Private Sector Productivity', *Economic Studies*, Vol. 17, pp.63–98.

Glimstedt, H. and Mariussen, A. (1998), *Introduction, Moving Beyond Convergency Theory, Nordic Institutions and Regional Development in a Globalised World*, Mariussen, A., Stockholm, Nordregio.

Gramlich, E. (1994),'Infrastructure Investment, A Review Essay', *Journal of Economic Literature*, Vol. 17, pp.1176–1196.

Hey, Ch., Pfeiffer, T. and Topan, A. (1996), *The Economic Impact of Motorways in the Peripheral Regions of the EU – A Literature Survey for the Royal Society for the Protection of Birds and Bird Life International*, Freiburg: EURES, Institute for Regional Studies in Europe.

Hook, W. (1996), *Counting on Cars, Counting Out People, A Critique of the World Bank's Economic Assessment Procedures for the Transport Sector and their Environmental Implications*, New York, Institute for Transportation and Development Policy.

Hook, W., Peters, D., Stoczkiewicz, M. and Suchorzewsky, W. (1999), *Transport Sector Investment Decision-Making in the Baltic Sea Region*, Berlin, German Federal Environmental Agency (Umweltbundesamt) (online under www.ITDP.org/pub.html).

Hurst, C., Thisse, J.F. *et al* (2000), *What diagnosis for Europe's Ailing Regions?* EIB Papers/ Cahiers BEI, Vol. 5, No. 2, pp.9–30.

Krugman, P. (1991), *Geography and Trade*, Cambridge, MA, MIT Press.

Martin, P. (1999), *Are European regional policies delivering?* EIB papers/Cahiers BEI, Vol. 4, No. 2, pp.11–23.

Martin, P. (2000), *The Role of Public Policy in the Process of Regional Convergence*, EIB Papers/Cahiers BEI, Vol. 5, No. 2, pp.69–80.

Martin, R. (1999), *The Regional Dimension in European Public Policy, Convergence or Divergence*, London and New York, MacMillan and St Martin's Press.

Mas, M. *et al.* (1996), 'Infrastructure and Productivity in the Spanish Regions', *Regional Studies*, Vol. 7, No. 30, pp.641–9.

Maystadt, P. (2000), *Preface [Special issue Regional development in Europe, An Assessment of Policy Strategies]*, EIB Papers/Cahiers, BEI, Vol. 5, No. 1, pp.4–5.

Mikulic, N. and Dusik, J. (1999), *Recent Developments in SEA in CEE*, OECD/ECMT Conference on SEA for Transport, Warsaw, 14–15 October.

Moucque, D. (2000), *A survey of Socio-Economic Disparities between the Regions of the EU*, EIB Papers/Cahiers, BEI, Vol. 5, No. 2, pp.13–24.

NSPA (2000), 'National Spatial Planning Agency, Dutch Ministry of Housing, Spatial Planning and the Environment', *Spatial Perspectives in Europe*, The Hague, NSPA.

Pinder, D. (1983), *Regional Economic Development and Policy, Theory and Practice in the European Community*, London, Allen G. and Unwin.

Quehenberger, M. (2000), *Ten Years After, Eastern Germany's Convergence at a halt?* EIB Papers/Cahiers, BEI, Vol. 5, No. 1, pp.117–136.

Regional Environmental Centre for CEE (REC), Sofia EIA Initiative (2000), *SEA of VI. Multi-Modal Transport Corridor Warsaw-Budapest, Scoping of Environmental Issues*, Regional Synthesis Report, Final Draft, Szentendre, REC.

Richardson, T. (1997), 'The Trans-European Transport Network: Environmental Policy Integration in the European Union', *European Urban and Regional Studies*, Vol. 4, No. 4, pp.333–346.

Rosset, B. (2000), *Contributing to regional development through project selection*, EIB Papers/Cahiers, BEI, Vol. 5, No. 1, pp.137–148.

Rzeszot, U., *et al.* (2000), *SEA of Multimodal Transport Corridor Warzawa-Budapeszt Initial Scoping*, In-Country Scoping Report, Poland, Szentendre, REC.

Seitz, H and Licht, G. (1995), 'The Impact of Public Infrastructure Capital on Regional Manufacturing Production Cost', *Regional Studies*, Vol. 3, No. 29, pp.231–40.

Standing Advisory Committee on Trunk Road Assessment (SACTRA) (1999), *Transport Investment, Transport Intensity and Economic Growth*, London, UK Government Publications and http://www.detr.gov.uk/heta/sactra.

Teufel, D. (1989), *Gesellschaftliche Kosten des Strassengueterverkehrs (The Social Costs of Road Freight Transport)*, UPI Bericht Nr. 14, Heidelberg, Umwelt- und Prognose-Institut Heidelberg (UPI).

Thisse, J. F. (2000), *Agglomeration and Regional Imbalance, Why? And is it bad?* EIB Papers/Cahiers, BEI, Vol. 5, No. 2, pp.47–68.

TINA Secretariat (1999), *TINA Final Report*, Vienna, TINA Secretariat.

Turro, M. (1999), *Going Trans-European, Planning and Financing Transport Networks for Europe*, Oxford, Pergamon.

Vanhoudt, P., Mathae, T. and Smid, B. (2000), *How productive are capital investments in Europe?*, EIB Papers/Cahiers, BEI, Vol. 5, No. 2, pp.81–105.

Vickerman, R.K. (ed.) (1991), *Infrastructure and Regional Development*, Pion, London.

Vickerman, R.K., Spiekermann, K. and Wegener, M. (1999), 'Accessibility and Economic Development in Europe', *Regional Studies*, Vol. 33, No.1, pp.1–15.

Vickerman, R.W. (1995), '*Regional impacts of trans-European networks*', *Annals of Regional Science, an International Journal of Urban, Regional and Environmental Research and Policy*, Vol. 29, No. 2, pp.237–54.

World Bank (1996), *Sustainable Transport, Priorities for Policy Reform*, World Bank, Washington D.C.

Chapter 4

Norwegian Urban Road Tolling – What Role for Evaluation?

Odd I. Larsen

The Norwegian Toll Cordons

Toll cordons are operated in three Norwegian cities (Larsen, 1995). It started in 1985 when the City Council in Bergen decided to introduce a toll cordon in order to raise money for road projects that were badly needed. Under normal circumstances these projects should be financed from national road funds, but waiting for ordinary government funding would have caused considerable delay and improving the road system in Bergen was considered to be an urgent task. The proposed tolling scheme had to be approved by the parliament as all other toll financed road projects in Norway. An agreement was also reached with the Ministry of Transport that secured additional government funds that matched the revenue from the toll system and this also provided an additional incentive for the City Council. The toll system in Bergen came into operation in March 1996. According to the original plans it should be closed down in 2001.

In 1990 Oslo and Trondheim followed the example of Bergen and implemented similar, but technically more advanced, schemes that rely on electronic tolling. These are also intended as temporary measures to be operated for 15 years. These two cities also reached an agreement with the government that secured additional funding that should match the revenue from the toll cordons.

The revenue from the toll cordons is earmarked for road projects with a minor share going to infrastructure for public transport, mainly bus lanes. The limited allocation to public transport is partly due to the Road Act which explicitly states that road tolls can only be used for road projects. However, in Oslo, some of the ordinary 'road funds' has instead been allocated for the urban rail system to circumvent this legal restriction. Before they started the political and administrative process necessary to implement the toll cordons, the three cities had compiled a list of projects to be financed. In Bergen and Trondheim the projects were located within the cities, while in Oslo the projects to be financed were located both within the municipality of Oslo and in the county of Akershus which surrounds Oslo.

It is important to be aware of the fact that the only purpose of the toll cordons has been to raise money locally for projects where the financing actually is the responsibility of the central government. The fact that the toll cordons also can be used as an instrument in traffic management ('road pricing') was never an issue in the initial stages.

While toll financing of motorways or specific projects like road tunnels and bridges are common both in Norway and many other countries, the Norwegian urban toll cordons are quite unique. An important feature of these schemes is that tolling is done in places that may be far removed from the actual physical projects that are financed by toll revenues. Also, the toll cordons are located so that the impact on the motorists' route choice is very limited or non-existent. This is clearly an advantage compared to traditional toll financing where tolling of the users of new projects can cause adverse effects on route choice that may severely reduce the benefits of new road links. Having a broad revenue base in terms of traffic volume also means that toll rates can be kept fairly low compared with the toll rates that would be necessary if the same revenue should be raised by tolling the users of the actual projects.

The Norwegian toll cordons can be evaluated from different perspectives:

- the social profitability of projects planned and financed by toll revenues;
- the toll cordons as a means of raising revenue;
- the location of toll cordons and the pricing scheme;
- the political process and public attitudes and acceptance.

Evaluation must in any case be relative to some realistic alternative and take into consideration the financial and political constraints faced by the local authorities.

Was it Worthwhile to Improve the Road Systems?

The first question we may ask is whether the projects to be financed were desirable. This should be a minimum requirement. It does not make sense to introduce a controversial financing scheme in order to finance road projects that are not socially profitable in the sense that benefits exceeded costs.

What can be said on this issue is that a formal assessment of the projects, prior to implementation of the toll cordons, was not carried out on a full scale. At least in the Oslo-region it was also explicitly stated that the compiled list of projects was tentative and that each project should undergo a more thorough evaluation before a final decision was made on implementation. In the final stages of preparation for political approval of the tolling scheme in the Oslo-region, the issue of social profitability of the total 'package' was actually raised and the Institute of Transport Economics was commissioned to make the assessment. Time and resources did not allow a thorough evaluation, but the conclusion that emerged was that total benefits exceeded the total cost for the whole 'package'. However, nothing could at that stage be said about the individual items.

Later analysis showed that there was strong interdependence between the benefits of some of the major projects in Oslo, thus that the phasing of the projects should preferably be based on a comprehensive analysis in order to maximise discounted net benefits. Such interdependencies are to be expected in any road system with multiple bottlenecks. Actually a more pragmatic approach had to be used when it came to the phasing of projects. The fact that projects initially were in different stages of preparation meant that projects with plans that had been approved often gained priority.

However, despite the lack of a more comprehensive and formal assessment and an optimum phasing, it is fair to say that the road system in all three cities needed to be improved. Some very severe bottlenecks were clearly evident. These had emerged after three decades of strong growth in road traffic and only moderate investment in the main roads system.

This state of affairs must be understood in the light of the strong rural bias that exists in the Norwegian Parliament. While investments in national roads had high priority during this period, the larger urban areas were clearly lagging behind and this situation was bound to continue. The urban toll cordons can therefore to a great extent be seen as local efforts to overcome the national priorities with respect to the geographical distribution of road investments.

If national road funds had been allocated to projects according to the priority indicated by the benefit-cost ratio, the urban toll cordons would probably never have materialised.

Urban Toll Cordons as a Means of Raising Revenue

If a road system can be improved and national financing is not forthcoming despite clear evidence that the improvements will be socially profitable, it will be natural to look for other options with respect to financing. However, in most cases there will be a social cost of financing. With regard to the tax system in general, economists usually point to the marginal cost of public funds, which is a dead-weight loss imposed on the economy when additional public funds are raised by distorting taxes.

For traditional toll financing of road projects we have a cost related to the system of toll collection (CT) and also a change in consumers' surplus (ΔCS), as tolling usually means that road users pay a toll that grossly exceeds the marginal cost of using the facilities. If we use the term R to refer to toll revenue, the cost of financing can be defined as:

$$fc = \frac{CT - \Delta CS}{R - CT}$$

i.e. *fc* is the social cost of tolling per unit of net revenue.

For traditional toll financing we usually have $\Delta CS<0$, i.e. some costs must be added to the cost of collecting the toll due to the adverse effects on demand. On the other hand, if motorists initially pay less than the marginal costs for road use, as the case will be in congested systems without road pricing, tolling can actually mean that we get a more 'correct' pricing of road traffic on balance. This will imply that $\Delta CS>0$ and the magnitudes involved may also mean that $fc < 0$. In the latter case it is actually socially profitable to introduce tolling as 'road pricing' irrespective of any need for financing.

From a purely economic point of view, toll financing should be used for road projects if

fc < marginal cost of public funds

The 'official' estimate of the marginal cost of public funds in Norway has recently been put at 0.2. For all three cities it seems that *fc* < *0.2*. For Oslo, estimates also show that $\Delta CS>0$, while in Bergen and Trondheim the charges in consumers' surplus are very small and may even be negative due to very moderate congestion.

Thus it seems that even though no formal assessment was carried out prior to implementation, we can conclude that the urban toll cordons in Norway raises revenue at an acceptable cost compared to general taxation. There is hardly any doubt that *fc* would be much higher with traditional tolling on each new road facility. However, it should be possible to improve on the present systems, and especially the system in Oslo.

The Design of the Cordons

The common factors for the three toll cordons are that the cordon surrounds the central area of the city and that only inbound traffic is tolled. All three toll cordons are also intended to be operated for a period of 15 years.

The main characteristics of the three schemes are summarised in Table 4.1.

With regard to the location of toll gates there are some differences between the cities. The location of the gates in Bergen utilises the special topography in Bergen. The central business district (CBD) is situated on a small peninsula that is linked to the rest of the city by three bridges and two main roads on which toll gates are located. Bergen had previously a tolled tunnel. The existing toll gate on the tunnel was incorporated in the cordon. The location of the toll gates also means that it is mainly traffic going to or through the CBD area that is tolled although a few other traffic flows are also affected. The topography also made it possible to design a cordon with a small number of gates relative to the size of the city and the location of the gates is in a sense very natural.

In Oslo the toll gates are generally located further from the CBD area in three distinct corridors. Except for this fact, it is difficult to point to any general principle that govern the location of the toll gates in Oslo. The distance from the CBD area to the toll gates varies between three and eight kilometres. It is probably fair to say that the location of the toll gates in Oslo is the result of a compromise between several objectives, partly political and partly related to practical aspects (costs, revenues and ownership of roadside land that could be used for the toll plazas). In a few places the toll cordon also cuts across local neighbourhoods.

The toll cordon in Trondheim is also located some distance from the central area. Even though Trondheim only has half the population of Bergen, it was necessary to have 12 toll gates. Like in Oslo the cordon in some places divides up local neighbourhoods. The location of some of the toll gates in Trondheim has recently been changed.

Table 4.1 Main Characteristics of the Toll Cordons

	BERGEN	**OSLO**	**TRONDHEIM**
Established	January 1986	February 1990	October 1991
Number of gates[1)]	6 (+1)	18 (+1)	12
Period of operation	Mon – Fri 6 AM – 10 PM	Continuously	Mon–Fri 6 AM–5 PM
Mode of operation	Manual (booths with attendants and reserved lanes for users with seasonal pass)	Semi automatic (reserved lanes for user with electronic tag, lanes with coin machines and lane with attended toll booth)	Automatic (reserved lanes for users with electronic tag, lanes with ticket vending machine and 2 gates have also a lane with attended toll booth)
Toll rates (NOK light/heavy vehicles)[2)]	5/10 with discount for seasonal passes and prepaid trips	10/20 with discount for seasonal passes and prepaid trips[3)]	max 10/20 with discount for prepaid trips. Time of day variations in rates.

1) One gate was later added to the systems in Bergen and Oslo. 2) 1 NOK ≈ 0,125 EURO; heavy vehicles are defined as vehicles with allowed payload exceeding 3500 kilograms. 3) At present 12/24.

In all three cities the toll system is operated by a public company that transfers the net revenue to the road authorities.

Bergen has a completely manual scheme. At all gates there are reserved lanes for vehicles with a seasonal pass (monthly, biannual or annual). The users of seasonal passes have a sticker on the windscreen and do not stop at the gates. The reserved lanes are subject to random controls and violators (i.e. vehicles without a valid seasonal pass) are fined. Supplementing the reserved lanes are lanes with toll booths. Coin machines are not used in Bergen.

Initially vehicles with a seasonal pass constituted 58–59 per cent of the tolled trips. This percentage has slightly increased over time and is now 62–63 per cent. The toll rates remained constant from 1986 through 1999, i.e. they decreased in real terms. In 2000 the rates were doubled.

Oslo started out with a system that was nearly identical to the one in Bergen (i.e. reserved lanes for vehicles with seasonal pass, at least one lane with a manned toll booth at each gate, but also lanes with coin machines). In November 1990, Oslo switched to an electronic seasonal pass that greatly improved the control of the lanes, reserved for users of a seasonal pass. The sticker on the windscreen was replaced by an electronic tag that is detected at the toll gates. If a vehicle without a tag uses the reserved lanes, the license plate is photographed and the car owner is fined.

In December 1991, the electronic system was extended to include prepaid trips. The electronic tag is issued to people who want to use the system with prepaid trips. They have to pay a deposit for the tag and an amount to cover a certain number of trips. Discounts are offered for prepaid trips and the rate of discount depends on the number of prepaid trips. Each trip is deducted from the users account for prepaid trips. Parallel to the introduction of prepaid trips the toll rate was increased by one NOK for manual payment. This was mainly done to promote the electronic scheme.

Trondheim started out in October 1991 with a more sophisticated version of the same electronic system that is used in Oslo. Only two of the toll gates have manned toll booths. At all gates it is possible to buy tickets from vending machines and for drivers to communicate with a control centre. Seasonal passes are not used in Trondheim. There is a choice between pre-payment and post-payment for the users of the electronic system. The discounts offered depend on whether it is pre-payment or post-payment and on the amount prepaid. There is a maximum limit on the number of paid trips per month (75 trips) and a limit of one tolled trip per hour. If the number of trips exceeds these limits, additional trips are not added to the bill, or deducted from the account. For the 'tag-holders' the rate is differentiated by time of day. Between 6 AM and 10 AM the toll rate varies between 8 NOK and 6 NOK depending on the amount prepaid, and after 10 AM the rate varies between 6 NOK and 4 NOK. Single tickets have a flat rate of 10 NOK.

It is somewhat paradoxical that Trondheim, being the smallest of the three cities and the least affected by congestion, has adopted a rate system that has at least some resemblance to congestion pricing. Oslo, on the other hand, where congestion pricing really might have something to offer, has a system that is operated around the clock seven days a week with no differentiation of rates by time of day.

The electronic system used in Oslo and Trondheim is based on a passive tag. This means that a central computer handles all transactions. Due to the issue of privacy, restrictions are placed on the amount of information that is permanently stored. The tag is assigned to the vehicle and car owners are not allowed to move the tag between different vehicles owned by a household. The same applies to the stickers used for seasonal passes in Bergen. The toll cordons in Oslo and Trondheim are the first implementations of automatic vehicle detection and debiting in urban areas.

Public Attitudes

A toll cordon, or for that matter any scheme designed to extract money from the motorist, is for obvious reasons not a popular measure. In Norway this is to some extent amplified by the fact that taxes on petrol and car ownership are already very high. The government also spends less on roads than the revenue from taxes on car ownership and car use. Consequently the taxes have a strong fiscal element. This is reasonable if the taxes on fuel and vehicles are also meant to cover an appropriate share of environmental and accident costs. Still, car owners tend in general to be against toll cordons.

There are probably four important reasons why it was possible to implement the toll cordons more or less against the public opinion:

- the need to improve the road systems and the lack of funds for this purpose became very obvious to the local politicians;
- the major parties represented in the City Councils agreed that the implementation of a toll cordon scheme should not be made into a major political issue;
- the revenue should be used for road improvements, i.e. to the benefit of road users;
- the government was in favour of the local initiatives and supplemented toll revenue with additional funds, i.e. a central government 'carrot' was important.

The controversy was probably greatest in Oslo where the implementation of the toll cordon was nearly halted due to demands in the City Council that toll revenue should also be used for public transit. Within the restrictions imposed by the Road Act, toll revenues could only be used for separate bus lanes or other road-related improvements for public transit. The demands on contributions to public transit went beyond such measures and included the rail-based systems in Oslo. The government had to mediate on this issue and the solution was that more of the additional government funds allocated to Oslo should be dedicated to public transit projects while the toll revenue should be used for road-related projects. The net result, although not formally, was that the toll cordon increased the funds available for the rail based systems.

In Bergen an opinion poll taken in December 1985 showed that 54 per cent of the respondents were opposed to the use of road tolls to finance road improvements in Bergen. Only 13 per cent were unreservedly in favour. The remaining 33 per cent were either for or against with reservations or had no opinion on the issue. A similar poll taken one year later showed only 36.5 per cent against and 50 per cent in favour with 13.5 per cent indifferent or without an opinion.

Similar opinion polls in Oslo showed that 65 per cent of the respondents were opposed to the toll cordon in November 1989 and 60 per cent in November 1990. In Trondheim the percentage opposed to the scheme switched from 72 before to 48 after.

It thus would appear that in all three cities a majority of the population was opposed to the toll cordons prior to implementation. Oslo still had a majority that remained negative after the toll cordon had been operated for a year, but the change in Bergen and Trondheim is remarkable. Generally it seems that public opinion gradually shifted in favour of the toll schemes, but at least in the Oslo region there is still a majority opposed to the scheme.

A major reason for the shift in opinions is probably that people experienced less adverse effects than they had expected. The toll gates did not introduce additional bottlenecks in the road system and free flow conditions were maintained in the lanes for users with seasonal passes (Bergen) or electronic tag (Oslo and Trondheim). The use of seasonal passes and an upper limit on the number of charged trips per month (Trondheim) meant that the maximum amount paid in tolls also remained acceptable. The cost of a monthly pass for a light vehicle in Bergen and Oslo is, for example, substantially less than the cost of a monthly pass for public transit. The less pronounced shift in attitudes in Oslo may be related to the fact that the toll cordon in Oslo is operated continuously and thus affects social trips in the evenings and weekends to a larger extent than the cordons in Bergen and Trondheim.

The local branches of the National Roads Administration are responsible for the projects that shall be financed by toll revenue. They put up road signs at construction sites stating that 'this project is financed by the toll cordon'. People are thus also constantly reminded that the tolls have visible results in terms of improvements in the road systems and are not just another tax.

A group initially opposed to the toll scheme in all three cities was the retail traders inside the cordons and especially in the CBD – areas. However, even *à priori* it is not clear that the toll cordons will have a negative impact on retail trade in the city centres. We should expect that people living inside the cordons will tend to do more shopping inside the cordons and this will counteract the impact on customers from outside the cordons. Saturday is also the 'shopping day' in the CBD – areas and in Bergen and Trondheim there is no tolling on Saturdays. Anyway, it has not been possible to detect any effect in the volume of sales and the retail traders do not seem to worry about the toll cordons any more.

Toll Cordons as 'Road-Pricing'

The Norwegian urban toll cordons have often been referred to as 'road pricing'. If we broadly define 'road pricing' as cost-based charging[1] for road use, the toll cordons can not be classified as 'road pricing' and this has never been the intention. The Road Act clearly states that tolling can only be used as a means of financing road projects. Initially this was also interpreted to mean that the toll cordons could not be used as an instrument in traffic management. However, later on it has been clarified that there is nothing in the Road Act that actually prevents toll rates from being differentiated by time of day. This means that the toll cordons can be used for a somewhat primitive type of congestion pricing if this is decided locally. The present taxes on fuel and vehicles in Norway are so high that it will mainly be driving in congested conditions that is 'undercharged'.

However, by now the road systems in Bergen and Trondheim have been improved to the point where congestion pricing probably has little to offer in terms of improved efficiency. In Oslo congestion will persist in the morning and afternoon peak in the foreseeable future and both the City Council in Oslo and the County Council in Akershus have shown interest in the principle of differentiated toll rates. Recently the Institute of Transport Economics completed a study on the possible impact of differentiated toll rates. The study was commissioned by the administration in Oslo following a request by the majority in the City Council.

At present, the average toll for light vehicles is close to NOK 9 (≈1.1 EURO). Estimates of the marginal external congestion cost for different types of trips in the Oslo region shows that this cost is much higher than the toll rate for most trips that cross the cordon in peak periods. The toll rate is reasonably in line with the marginal cost for trips between the peaks, but exceeds the marginal external cost for periods with low traffic volumes, e.g. evenings and weekends.

This structure indicates that it should be possible to improve on the present situation by differentiating the toll rates by time of day. A disadvantage in this respect is that toll is charged only for inbound trips. If the present system is considered as a congestion pricing scheme we have to make allowance for this fact. The toll for inbound trips should preferably also include a component for the return trip. However, as the timing of the return trips is not known it would not be reasonable to charge as if both the inbound and the outbound trip takes place during the peak hours.

People will adapt to differentiated toll rates both by changing to public transport and by the timing of car trips and may also change destination. An impact assessment of differentiated toll rates cannot disregard the impacts on public transport in peak periods because parts of the public transport system are then used to capacity.

The approach taken in the study can briefly be described as follows (see Grue, *et al.*, 2000, for more details):

- we assumed that any change in toll rates (and public transport services) would only affect mode choice and the timing of car trips within the peaks (defined as 3 hours in the morning and in the afternoon);
- OD-matrices were calibrated for both road traffic and public transport. Each matrix represented a typical situation. For the morning and afternoon peaks we had separate matrices for the 'maximum' hour and the hours before and after. Separate matrices were used for an hour between peaks and for an average 'low traffic' hour. Multiplying the results by the appropriate number and adding up will then give results on an annual basis;
- for car matrices and public transport matrices respectively we assumed that a certain percentage was captive riders. For the remaining riders we assumed that a logit-model could adequately describe the choices (mode and timing within the peaks) and calibrated a model that reproduced the observed matrices with a sufficient degree of accuracy;
- a combined route choice and mode choice model was run with different toll rates and iterated to equilibrium. In order to get realistic results we also updated the frequency of public transport services between iterations. Updating was confined to lines that got excessive load factors.

This method will probably underestimate the long run impacts on demand due to the exclusion of impacts on destination choice in the model. On the other hand, by this method we attempt to model the change from one equilibrium to another which is very difficult if destination choice is handled simultaneously.

The alternatives analysed in this study, against a base case of an undifferentiated toll of NOK 9, were:

- *Alternative 1 (A1):* The toll is differentiated over the three hours in the morning and afternoon rush (NOK 20, 40 and 30 respectively). The toll in the mid day periods (incl. Saturdays 9–15) is assumed to have the same level as today (NOK 9). Between 6 p.m. and 6 a.m. on working days and in low traffic periods in weekends no toll is charged;

- *Alternative 2 (A2):* The toll is differentiated over the three hours in the morning and afternoon rush but with lower fees (NOK 15, 35 and 25 respectively). Otherwise the fees are the same as A1;
- *Alternative 3 (A3):* The toll is increased to the same amount in each of the three hours in the morning and afternoon rush (NOK 25). Otherwise the fees are the same as A1;
- *Alternative 4 (A4):* Same as A3 but with free travel on Saturdays as well.

The alternatives were evaluated in a benefit-cost framework and Table 4.2 shows the estimates of different items and the total. Two items not included in the table is the value of savings in fuel consumption for road traffic that remains in peak periods, and savings in operating cost for the toll cordon due to a shorter period of tolling.

Table 4.2 Benefits for the Alternative Systems (Million NOK/year)

	A1	A2	A3	A4
'consumers surplus'	13	102	90	135
toll revenues	209	117	120	76
fare revenue PT[a]	139	107	101	98
operating cost PT	-164	-151	-135	-136
capital cost PT	-24	-20	-18	-18
Sum	174	155	158	155

a) Public transport.

With inclusion of the missing items, the annual benefits will probably exceed 200 Mill NOK for alternative A1. The estimates in Table 4.2 shows that the transfer of trips to public transport in peak periods will increase the financial deficit of the public transport system. This only confirms that public transport fares in peak periods are below marginal cost for the operators.

It is reasonable to assume that consumer surplus will be reduced by introduction of road pricing. The reason for the positive impact in Table 4.2 is twofold:

- public transport services are improved due to increased frequencies and this produces benefits also for current public transport riders;
- tolling is dropped in 'low traffic' periods.

Table 4.2 does not include any benefits related to environmental improvements or improvements in traffic safety. The reason is that these benefits will be balanced, although not exactly, by a reduction in government revenue from taxes on fuel and vehicles.

This is not to say that environmental benefits will be negligible. The authorities in Oslo are very concerned with air pollution. Especially in the winter, the level of air pollution can be severe in the central area of Oslo, an area that also has a high population density. An assessment of impacts on air pollution was not carried out, but the model was used to estimate changes in fuel consumption that is correlated with air pollution. Table 4.3 shows that there is a substantial reduction in fuel consumption on roads inside the cordon in the rush hours. On an annual basis for the whole region, the impact is much smaller.

Table 4.3 Annual Amount of Fuel Consumption (Million Litres) in Different Parts of the Area in Different Periods and Changes Produced by the Alternative Systems (Million Litres Per Year)

Alternative	Total	In rush hours	Inside cordon	Inside cordon in rush hours
BASE	637	215	236	82
A1	-4.7%	-14.0%	-5.8%	-18.0%
A2	-3.7%	-11.3%	-4.6%	-14.5%
A3	-3.6%	-10.9%	-4.4%	-13.9%
A4	-3.5%	-10.9%	-4.2%	-13.9%

Despite the obvious limitations of the toll cordon, the study shows that it should be possible to differentiate the toll rates by time of day and thus get a fair share of the benefits that can be expected from more sophisticated congestion pricing.

From a practical point of view the main problems with differentiated rates in Oslo will be how to avoid great and sudden leaps in the toll rates. With the electronic system this is fairly simple to avoid, but coin machines and manned toll booths are not so easily adapted.

The study has so far not been presented to and discussed in the City Council and it is by no means sure that there will be a majority in the Council for differentiating the toll rates. After all, the City Council's request was for a study and a proposal, and the majority in the City Council did not make any commitment to the implementation of differentiated rates. Despite the fact that all the alternatives with differentiated roll rates show an improvement over the present situation in terms of net benefits, any change that makes a large group of motorists worse off is bound to be politically controversial. The group that will be worse off is the commuters that cross the cordon by car in peak periods.

Conclusions

The Norwegian urban toll cordons were initiated in order to finance what was considered a need for urgent improvements in the road systems of the three cities. At that stage only a few projects had actually been subject to formal evaluation in terms of costs and benefits. However, available evidence indicates that the 'packages' implemented were socially profitable even though they may contain some projects that are not socially profitable. On the other hand, investment plans of the magnitude involved in the Oslo region should warrant a more thorough evaluation. Even the phasing of different projects in time can have important consequences for the total benefits when an investment programme is undertaken that spans a period of 15 years.

Toll cordons as a means of raising funds for the investments have clear advantages in urban areas where traditional tolling can cause very adverse effects on route choice. It is also clear that much can be gained by a careful design of toll cordon schemes. In this respect, the Norwegian examples are not particularly well designed. Even though 'the cost of funds' with the present schemes compares favourable with the cost of raising revenue by general taxation, there is certainly much to be gained by a better rate structure and probably also a better location of toll gates in the Oslo region. In congested urban areas the potential for adapting toll rates to the principles of 'congestion pricing' can be utilised to further reduce the 'cost of funds' and it may also be worthwhile to consider tolling in both directions and extending the system.

Note

1 'Cost-based' must here be understood as short run marginal cost, i.e. the costs that can be attributed to the individual trips.

References

Grue, B., Hamre, T., Rekdal, J. and Odeck, J. (2000), 'From Cordon Toll to Congestion Pricing in Oslo – What are the Benefits?', Paper presented at CODATU, Mexico City.

Larsen, O. I. (1995), 'The Toll Cordons in Norway – An Overview', *Journal of Transport Geography*, Vol. 3, No. 3, pp.187–198.

Ramjerdi, F. and Larsen O.I. (1991), 'Road Pricing as a Means of Financing Investments in Transport Infrastructure – The Case of Oslo', Paper presented at PTRC, Sussex.

PART II
TECHNICAL ASPECTS OF EVALUATION

Chapter 5

Spatial Economic Impacts of Transport Infrastructure Investments

Jan Oosterhaven and Thijs Knaap[1]

Summary

Estimating the spatial economic impacts of transport infrastructure is an unsolved issue that has plagued economic science for a long time. This paper will give an overview of the basic problem, the contributions of the empirical literature, the modelling approaches used until now, a recent application, and a perspective on how to proceed.

The basic problem lies in establishing the '*anti monde*', that is, the economic development that would have occurred without the investment in infrastructure. This basic problem is burdened with uncertainty about the direction of the impact of new transport infrastructure on the regions or nations affected. As infrastructure reduces the cost of both imports and exports of goods and services, the net effect is not clear.

Macro economic research only gives indications about the demand and supply effects of bundles of historical investments in infrastructure. This only with considerable debate about the econometrics and the causality implied by the estimations. Moreover, macro research has only limited value when taking decisions on specific infrastructure projects. Surveys among firms and consumers have various measurement problems, but they also have the advantage of providing *ex ante* micro information. Among the major disadvantages are strategic answers, sample selection and the inability to capture indirect effects on non-using firms or consumers.

Economic potential models for interregional infrastructure and land-use/ transportation interaction (LUTI) models for intra-urban infrastructure provide the best empirical answers to the question of the economic impacts of transport investments. Both approaches, however, show one major defect, namely an unsatisfactory foundation in economic theory. A promising alternative is provided by the new economic geography models that are evolving into more broadly based spatial computable general equilibrium models (SCGE models).

Results of a recent Dutch application of a new SCGE model with 14 sectors and 548 municipalities on a proposed transrapid (magnetic levitation train) project from Schiphol Airport to Groningen City, confirm that the SCGE approach has a high potential. Moreover, its implementation appeared to be far easier than was expected. Some aspects and outcomes of this recent study will be discussed in more detail.

Despite these difficulties, empirical literature has produced some general qualitative outcomes. First and foremost, both SCGE models and potential models show the same spatial pattern of impacts of infrastructure. On an isomorphic plane, investments in line infrastructure (road, rail, air and waterways) produce butterfly type of spatial patterns with diminishing impacts as the butterflies grow larger. Improvements in point infrastructure (terminals and harbours) produce concentric ring impact patterns.

Second, there appears to be general agreement on the fact that new infrastructure produces minimal impacts in countries with abundant infrastructure services. This holds with one major exception. When new infrastructure resolves strong capacity limits in either point or line infrastructure, the local effects will be considerable but mostly at the expense of cities and regions close by.

Introduction

When considering the economic effects of transport infrastructure, first, one has to make a distinction between direct and indirect effects, temporary and permanent effects, and market and non-market or external effects (see Table 5.1).

Temporary economic effects will occur during construction, directly and indirectly through demand effects. Less discussed, but equally important, are indirect supply or crowding-out effects, both through the capital market as a consequence of the need for finance and through the labour market as a consequence of drawing on specific spatial and occupational segments of that market. Besides, there will be direct temporary external effects, such as noise and environmental disturbances during construction activities, and indirect temporary external effects, such as emissions due to backward economic effects (far) away from the actual construction sites.

Table 5.1 Type of Effects of Transport Infrastructure Investments

		Temporary:	Permanent
Direct	via markets: external effects:	Construction effects Environmental effects	Exploitation and time saving effects Environmental, safety etc. effects
Indirect	via demand: via supply: external effects:	Backward expenditure effects Crowding-out effects Indirect emissions	Backward expenditure effects Productivity and location effects Indirect emissions etc.

Permanent direct economic effects include exploitation costs as well as transport costs and time benefits for people and freight. These user benefits, mostly, are the prime reason for investing in infrastructure projects. When this is the case, one speaks of a *passive infrastructure policy*, meaning that investments primarily follow the growing demand for transportation where it occurs, and attempts to avoid or mediate the costs of congestion. Passive infrastructure policy decisions typically will be made only on the basis of a positive net national total of users' benefits minus investment, exploitation and external cost.

Besides direct effects, there will of course also be permanent indirect economic effects. First, these relate to the backward expenditure effects of the exploitation and use of infrastructure. Second, these relate to the so-called programme or induced effects, which are defined as the consequences of the reduction in transport cost for production and location decisions of people and firms, and the subsequent effects on income and employment of the population at large (Rietveld and Nijkamp, 2000). Naturally, these supply-driven effects will, in turn, also have demand effects.

When attaining these cost-induced effects is the main objective of investing in infrastructure, one speaks of an *active infrastructure policy* that tries to influence location and production decisions of firms and thus tries to induce private investments. Besides a hopefully positive net national total of benefits and cost, active infrastructure policy decisions will typically be motivated either by the pattern and size of the regional re-distributive effects and/or by the expected size of the national generation effect.

Besides these permanent economic effects there will also be permanent effects that are external to the market, such as noise, safety, emissions and environmental disturbances (Rothengatter, 2000). Moreover, the indirect economic effects also cause indirect external effects that need to be incorporated in the analysis when a fair valuation of investments in alternative transport systems is concerned (e.g. Bos, 1998, for the indirect energy use and emissions of different freight transport systems).

In this paper we will concentrate on the indirect effects of infrastructure on the regional and national economy (the cursive effects in Table 5.1). First, we will discuss the various methodological approaches available to tackle this complicated problem. Second, we will present some empirical results of a spatial CGE model for the Netherlands, which was recently built to estimate the indirect effect of proposals for new rail infrastructure between the West and the North of the country. Finally, we will summarise the main conclusions that may be drawn from the empirical work in this field.

Methods to Estimate the Economic Impacts of Infrastructure Investments

There is a large amount of literature on the economic impacts of infrastructure (Blonk; 1979, Vickerman, 1991; Rietveld and Bruinsma, 1998) as well as a large variety of methods to estimate these impacts (Oosterhaven *et al.*, 1998; Rietveld and Nijkamp, 2000). The methods most used are the following:

- micro surveys with firms;
- estimations of quasi production functions;
- partial equilibrium potential models;
- regional and macro economic models;
- land-use/transportation interaction (LUTI) models; and
- spatial computable general equilibrium (SCGE) models.

Below we will briefly review these approaches.

Micro Surveys with Firms

There is a rich theoretical literature on the influence of infrastructure and accessibility on the location decisions of firms. The past theoretical literature emphasises the importance of minimising transport cost (Weber, 1909), which is not surprising in view of the high absolute and relative transport cost of that time. Present literature tends to de-emphasise the importance of transportation cost as compared to other costs such as labour costs (Dicken, 1986). An interesting exception is McCann (1998) who replaces the concept of transportation cost with that of logistic costs and argues that the latter play a central role in locational decisions of most large multinational firms. Dutch empirical research seems to substantiate this (BCI/NEI, 1997). The large Dutch share in the total of all European Distribution Centres may be explained by its low transportation cost to the rest of Europe due to the Rotterdam harbour and Schiphol airport transport networks.

This literature, however, does not serve to answer the empirical question about the impacts of specific types of infrastructure in specific locations (Oosterhaven *et al.*, 1998). To this aim micro surveys with firms provide more answers. There are two strands within this line of research. First, one finds a series of general surveys with questions about the importance of all kinds of location factors including infrastructure accessibility. Naturally, the answers differ from country to country, as firms are confronted with different locational bottlenecks in different countries, and tend to bias their answers in the direction of those location factors that they want to see improved. The answers also differ according to the type of firms/sectors that are interviewed, as different sectors have different cost profiles and different market positions. The conclusion seems to be that centrality and the reliability of access are important, but not the actual transportation cost. In fact, for most sectors, Europe presents a rather level paying field within which secondary and even subjective factors of location play an increasing role (Pellenbarg, 1998).

The second strand of micro survey research tries to investigate the historical or future impacts of specific infrastructure investments. The outcome of this type of research is often rather dubious, since the purpose of the investigation is seldom hidden, and firms tend to answer positively even when the project at hand is of little importance to their own firm. Such strategic and socially desirable answers are very difficult to evade (NEI/TNO/RUG, 1999, for ways to circumvent some of these problems). Besides, when different variants of the same infrastructure are under investigation, a questionnaire approach becomes unwieldy. Finally, it should be noted that micro surveys do not indicate which firms (further away) are indirectly influenced by the actions of the directly affected firms (closer by).

Quasi Production Functions, Accessibility and Potential Models

Within macroeconomics the infrastructure debate started with the claim that the productivity decline in the USA was caused by a lack of investments in infrastructure (Aschauer, 1989). Since then a whole series of articles appeared that partly substantiated and more often weakened the original statement. The most common approach is the *quasi production function* approach:

$$Y_r^t = f(L_r^t, K_r^t, \mathit{Infrastructure\ stock}_r^t) \quad (1)$$

Besides labour (L) and (K) per region or nation per time period, several components of – or the total stock of – infrastructure is included in a macro production function in order to explain the level or change of domestic output (Y). This approach has several problems.

There are complicated econometric issues, relating to the one-sided nature of either the time series data (only data on different t's) or the cross section data (only data on different r's) that are used most often (see Sturm, 1998, for an overview).

The direction of causality is not easily detected statistically, as infrastructure may both follow and lead economic growth. To sort out the causality issue very long series of panel data (both t's and r's) are needed, but these are hardly available (for a recent Dutch attempt, see Van Ewijk, Hakfoort and Schnieders, 2000).

Measurements of the infrastructure *stock* fail to take account of the actual supply of the infrastructure services that determine its productivity contribution (for instance, infrastructure 'white elephants' are part of the stock but do not produce services). Historically found macro elasticities are of no use when the decision about specific individual projects has to be taken, as such projects are both specific in type (line, point, etc.) and specific in their location within the network.

As a result no clear conclusion has been reached. Macro production elasticities of infrastructure are found to vary considerably among the different studies. They sometimes have negative signs and often are not found to be significant (for an overview see Sturm, 1998). At the regional level, comparable attempts have been made, using several infrastructure stock indicators, but here the results are hardly more clear (Rietveld and Nijkamp, 2000).

When a detailed spatial division of the study area is used, the last two problems above may be remedied when some measure of the *economic accessibility* of region r is used in (1) instead of the stock of infrastructure (for overview, see Jones, 1981; Rietveld and Bruinsma, 1998):

$$\mathit{Accessibility}_r = \Sigma_s Y_s f(c_{rs}) \quad (2)$$

In (2) f is a downward sloping (gravity or preferably entropy) function of the communication cost between region r and region s (c_{rs}). The inverse of (2) gives the economically weighted average communication cost or distance of location r to the total study area. Obviously, (2) allows approximating the increase in economically useful infrastructure *services* available to a certain region that will result from investing in specific lines or nodes of the networks included. Moreover, (2) also shows that not only the region wherein the actual investment takes place will profit from improved accessibility. In fact, a whole series of regions/nations will profit from any investment as indicated by the summation and the distance function. The inability to deal with the multi-regional use of infrastructure is the fundamental flaw of the stock measures for individual regions in (1).

Using (2), the *economic potential* concept provides an approximation of the significance of changes in accessibility for the economy of the region at hand:

$$Potential_r = Y_r \Sigma_s Y_s f(c_{rs}) \qquad (3)$$

This almost directly follows from the fact that (3) is proportional to the total flow of traffic from region r (Wilson, 1974; Jones, 1981), which in its turn is proportional to the total size of the economy of region r. Evers and collaborators (Evers *et al.*,1987; Evers and Oosterhaven, 1988) were the first to use (3) to estimate the economic impacts of new infrastructure. They turned a variant of (3) with border dummies and a modal split parameter into a multi-sectoral potentials model, and used it to estimate the employment impacts of a proposed high speed rail connection from Amsterdam to Hamburg.[1]

Their approach was shown to have a micro economic (logit) foundation based on profit maximising locational behaviour of firms, and was shown to produce the 'right' spatial pattern of impacts but not necessarily the right macro level of these impacts (Rietveld, 1989). Later on Bröcker (1995) showed that the gravity type of a spatial impact pattern could also be produced by the even more satisfactory use of a spatial CGE model.

Regional and Macro Economic Models

Incorporating (2) in (1) provides a solution to the last two problems of the quasi production function approach. However, it does not solve the first two problems mentioned. To do that one needs a structural equation approach. The partial equilibrium potentials model using accessibility measures only provides a first step. The conceptual basis of a more comprehensive approach is given in Figure 5.1 (adapted from FNEI, 1984, and Rietveld and Nijkamp, 2000).

Figure 5.1 shows that all indirect economic impacts start from the supply side with transport cost and time gains. It further emphasises that new or improved infrastructure, in principle, may have both positive and negative economic effects for any region that is influenced by the consequent reduction in communication cost. For some sectors and products increased accessibility may boost that region's exports, whereas for other sectors and products it may lead to increased competition on its home market and a contraction of local output, income and employment. These positive and negative effects may well be enhanced because of economies of scale. When present, (internal) scale economies at the firm level will increase already positive impacts, whereas they may further the negative impacts for other sectors. These findings will be modified and complicated because of inter-industry and consumption demand feedbacks, which may lead to further (external) cluster economies for other not directly affected firms.

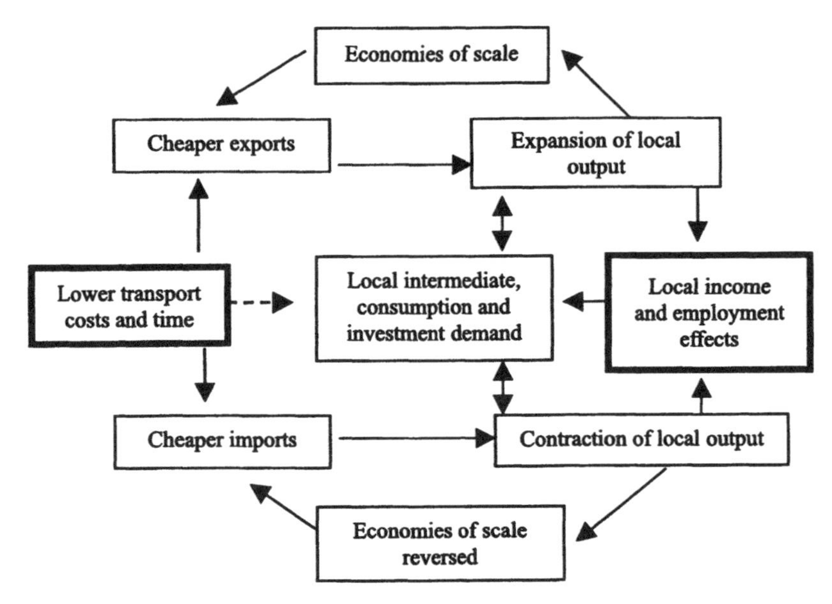

Figure 5.1 A Conceptual Model of Transport Infrastructure Impacts

Finally, the dotted line indicates the direct effect of (generalised) transport cost savings on the demand for all non-transport products. This indicates that the net regional or national welfare effect of new infrastructure tends to be positive, unless the contraction effects are really heavy and, of course, unless the project is too costly when compared to its benefits.

The possible crowding-out effects of the investment are not shown in Figure 5.1. Typically, macro economic models are well suited to model these effects (Van de Klundert, 1993; Toen-Gout and Van Sinderen, 1995; Eijgenraam, 1995). The question, however, is whether they may also be used to model the impacts that are shown in Figure 5.1.

As follows from Figure 5.1, regional and national economic models need to be multi-sectoral in order to capture the sectorally different nature of the primary impacts of infrastructure. Also the exogenous cost reduction impulse needs to be calculated in such a way that the sectorally different impact on import, export and domestic prices will be captured. Hence, a detailed transportation module is needed to generate this information. But even then, regional and national economic models will suffer from the fact that they do not have a spatial dimension, except for the presence of imports and exports to the rest of the world. Consequently, they will have great difficulty to capture the differences in the impacts when different locations of the same infrastructure investments are concerned.

LUTI and SCGE Models

Spatially detailed models provide the only way to adequately model the economic impact of new transport infrastructure. Here we will discuss two broad classes of such models, namely land-use/transportation interaction (LUTI) models and spatial computable general equilibrium (SCGE) models.

LUTI models consist of linked transportation models and 'land-use' or better location models. They mostly employ a *system dynamics* type of modelling and are primarily developed to predict future growth and to analyse policy scenarios for large urban conglomerations (e.g. Lee *et al*, 1995). There is a whole series of such models for different conglomerations. LUTI models have a decades long history of gradual development and are nowadays typically very disaggregated with numerous spatial zones, sectors, household types, transport motives, modes of transportation, etc. (DSC/ME&P, 1998; Wilson, 1998).

SCGE models typically are comparative static equilibrium models of interregional trade and location based in *microeconomics*, using utility and production functions with substitution between inputs. Firms often operate under economies of scale in markets with monopolistic competition of the Dixit-Stiglitz (1977) type. One of the few empirical applications of this approach is found in Venables and Gasiorek (1996). Interesting theoretical simulations with a SCGE model with a land market are found in Fan *et al.* (1998). A recent Dutch application will be discussed in the next section. These models are part of the new economic geography school (Krugman, 1991; Fujita, Krugman and Venables, 1999) and have been around for less than a decade. In other words, we are comparing a mature methodology, possibly at the end of its life cycle, and a new methodology that is still in its infancy.

The practical feasibility of LUTI models is large, which for a mature methodology with heavy investments in its empirical implementation is not surprising. Especially the transportation sub-models are known to be rather adequate in estimating all kinds of transportation price and quantity impacts of policy measures in the transportation sector itself. Given the scientific uncertainty around the location behaviour of firms, this does not hold to the same degree for the impact of transport measures on the location of firms. In view of the decrease of the relative cost of freight transportation over time, this is not too surprising. The relative cost of passenger transportation, however, has been increasing over time, mainly because of the increase in time costs due to increases in congestion and increases in real income. For this reason, the location of service activities can be explained much better than the location of industrial activities. As the location of most service activities primarily follows that of people and industrial activities, its location choices mainly play a role on the intra-urban level. Consequently, the power of LUTI models in estimating the interregional location effect of transport measures is much less than that of estimating the impact on intra-urban location decisions.

Finally, most LUTI models are not well able to translate the impacts of transport and infrastructure measures into estimates of consumer benefits, as is needed in a sound, welfare theoretically underpinned cost-benefit analysis (CBA). Whether this is the case, mainly depends on how consumer and producer behaviour is modelled and estimated. In the best LUTI models consumer choices relating to transportation and location decisions are usually modelled and estimated by means of a discrete random utility approach. Producer location decisions, however, are seldom modelled by means of discrete profit maximising behaviour, whereas producer production and price decisions are practically always modelled using some kind of fixed ratios. As a consequence, most LUTI models will provide reasonable estimates of direct transport user benefits. Sometimes they will also provide reasonable estimates of consumer benefits in as far as the latter are based on discrete choice behaviour. The existing LUTI models, however, are not able to estimate transport benefits that are based on continuous consumer choices or discrete and continuous producer choices.

SCGE models, typically, are theoretically well suited for this evaluation task (Venables and Gasiorek, 1998). The SCGE modelling problem, at the moment, is not theoretical in nature, but empirical and computational. The consistent estimation of all the necessary consumers' and producers' substitution elasticities is problematic, if only because of the lack of adequate data and the lack of a tradition of estimating such elasticities at the regional level. Moreover, the calibration of these models such that they reproduce recent history and simultaneously provide plausible (i.e. stable) projections is problematic too, especially because of the highly non-linear character of the behavioural equations.

Whether LUTI models can easily incorporate imperfect markets, and internal and external economies or diseconomies of scale, is doubtful. The strength of most LUTI models lies in their segmentation and detail, i.e. they usually contain many different zones, transport modes, households type, firms type, and so on. The benefit of having such detail lies in the homogeneity of behaviour and the assumed stability of relations at that level of detail. But this detail is achieved at the cost of mathematical and theoretical simplicity, such as perfect competition, fixed ratios, linear relations and the absence of scale economies.

The existing, still young SCGE models have opposite properties, namely a lack of detail or sound empirical foundation, but a sophisticated theoretical foundation and rather complex, non-linear mathematics. The latter is precisely the reason why SCGE models are able to model (dis)economies of scale, external economies of spatial clusters of activity, continuous substitution between capital, labour, energy and material inputs in the case of firms, and between different consumption goods in the case of households. Moreover, monopolistic competition of the Dixit-Stiglitz type allows for heterogeneous products implying variety, and therefore allows for cross-hauling of close substitutes between regions. Finally, SCGE models lead to a direct estimation of especially the non-transport benefits of new infrastructure, which are absent in most LUTI models.

Whether a further piecemeal improvement of the theoretically handicapped, but in practice successful LUTI models is preferable to the implementation of a theoretically superior, but as yet untested alternative, is essentially a matter of taste and belief. DSC/ME&P (1998) confess to the piecemeal improvement strategy. The further segmentation they call for may be necessary for the 'best' estimation of the impacts of transport policies, but it is not sufficient for the 'best' estimation of the indirect transport benefits needed for CBA. The latter requires modelling, not only of discrete choice, but also of continuous responses of consumers and producers based on, respectively, utility maximising and profit maximising assumptions. Our experience in model building tells us that the introduction of additional causal mechanisms or additional actors produces much more differences in the outcomes than a further differentiation of already present relationships and actors. We would rather like to stress the potentially dead-end character of that approach, and would like to advocate the more promising but also more risky start of empirically based SCGE modelling.

Two problems, however, remain. LUTI models are inherently more dynamic than the comparative static SCGE models. The latter, for the moment, are only able to compare the outcomes of different equilibrium states, such as:

- benefits of generalised transport cost reductions due to changing prices, production, consumption and trade, while holding the number of firms and the number of workers per region constant; showing what could be labelled as the *short-run* effects;
- benefits of transport cost reductions when the number of firms per region is allowed to change showing *medium term* effects;
- benefits when the number of workers is allowed to change too; showing the *long-run* effects of new transport infrastructure.

A truly dynamic SCGE approach is theoretically possible but raises a whole new series of issues (Knaap, 2000).

To some, it is not the comparative static but the equilibrium character of SCGE models that poses the fundamental problem. But this seems to be a less serious one as SCGE models may well incorporate disequilibrium features, e.g. (regional) unemployment caused by (nationally) set inflexible wages (Van den Berg, 1999, and the next section). In fact, solving the highly non-linear SCGE models becomes much simpler numerically when all kinds of prices, quantities and ratios are fixed, as is frequently done in the LUTI models.

Results of a Spatial CGE Model to Evaluate Dutch Rail Proposals

Some of the capabilities and problems of SCGE models may be illustrated with the recent construction and application of such a model to evaluate six alternative Dutch rail projects (Elhorst *et al.*, 2000 for the whole study).

These projects all connect the relatively rural Northern Netherlands (containing the provinces of Groningen, Friesland and Drenthe) with the heavily urbanised Randstad (containing the cities of Rotterdam, the Hague, Amsterdam and Utrecht, and their connecting surroundings). The faster rail connection should run through the new polder province of Flevoland, which used to be part of the former Zuiderzee. At present the rail connection between Groningen City and Schiphol Airport runs around the former Zuiderzee. Hence, rail travel distances and time will be shortened considerably, but car distances and time will hardly change as even a doubling of the modal share of rail will reduce the modal share of cars only slightly. The six different rail alternatives are summarised in Table 5.2.

Table 5.2 Description of the Rail Variants with their Travel Times Groningen-Schiphol (in minutes)

REF	Reference scenario or zero-alternative, including the so-called Hanze-line from Lelystad to Zwolle, which is not yet constructed	118
HIC	Hanze-line intercity, the only difference with REF is its higher speed	102
HHS	Hanze-line high speed, including rail shortcuts and a TGV only calling at major stations	71
ZIC	Zuiderzee-line intercity, a new rail track from Lelystad to Groningen with conventional trains	89
ZHS	Zuiderzee-line high speed, as ZIC but including a TGV only calling at three major stations	65
MZM	Trans-rapid metro, a new magnetic levitation rail system from Schiphol to Groningen calling at all 6 intermediate stations 6 times per hour	59
MZB	Trans-rapid high speed, as MZM but calling at the 6 stations 2 times, at the 3 major stations 2 times, and running a shuttle 2 times from Schiphol to Almere in Flevoland only	45

For details on the Dutch SCGE model (RAEM) model we refer to Knaap and Oosterhaven (2000). Here we will discuss only some of its most salient features and some of the problems in its implementation and application.

RAEM models demand, output and trade of 14 industries in 548 municipalities with one (presently) immobile household sector. All markets are of the monopolistic competition type and each firm in each industry produces one and only one variety of the product of that industry. In all production and utility functions the varieties x_i are added to Q_j with the following CES-function (Dixit and Stiglitz, 1977):

$$Q_j = \left(\sum_{i=1}^{n} x_i^{1-1/\sigma} \right)^{1/(1-1/\sigma)} \quad (4)$$

In (4) σ represents the elasticity of substitution among the n different varieties of industry j.

All utility and production functions have a Cobb-Douglas specification. The production function only uses intermediate inputs and labour:

$$Y_j = L_j^{\alpha} \left(\prod_{i=1}^{m} Q_i^{\gamma_{ij}} \right)^{1-\alpha} \quad (5)$$

In (5) α gives the substitution elasticity between labour and the total of the intermediate inputs and gives the substitution elasticities among the intermediate inputs from different sectors.

In the equilibrium all prices are a function of all other prices. In this solution the complement of the quantity aggregate (4) is the following price index function:

$$G_j(p_{ij},\ldots,p_{nj}) = \left[\sum_{i=1}^{n} p_{ij}^{1-\sigma}\right]^{1/(1-\sigma)} \tag{6}$$

In (6) p_{ij} is the price of variety i delivered to sector/consumer j. These purchasing prices are, of course, inclusive of the transport and communication cost of delivering variety i to sector/consumer j.

The way in which transport costs are included in the prices is decisive for the functioning of this type of model. In the monopolistic competition model the equilibrium mark-up, including the transport cost mark-up, is such that price is equal to average cost but larger than marginal cost. In view of the problem at hand, RAEM uses a new bi-modal (freight/people) transport cost mark-up:

$$p^{*} = [f_g(d_g)]^{\pi} \cdot [f_p(d_p)]^{1-\pi} \cdot p \tag{7}$$

In (7) π gives the importance of freight (g) transport for the transportation cost of the sector at hand. Information on this parameter proved to be scarce. Hence, expert judgement was used to 'guestimate' the 14 sectoral π's needed. In (7) f follows the usual specification of *iceberg* transport cost (e.g. Bröcker, 1998):

$$f(d) = 1 + \upsilon \cdot d^{\omega} \tag{8}$$

In (8) υ and ω are parameters to be estimated and d are the distances. For freight, simple road kilometres are used as freight distances do not change in the application. For people, transport cost varies between the infrastructure variants (see Table 5.2). The travel times used are weighted averages between times for cars, slow traffic (bikes etc.) and public transport. In the simulations only the last type of travel times changed, along with the modal shares. Both types of changes were derived from a very detailed transport model (LMS, see NEI, 2000, and Elhorst *et al.*, 2000).

The estimation and calibration of RAEM proved to be complicated, a situation that is quite common with both spatial and non-spatial CGE models. In fact, three types of parameters were used:

- parameters 'guestimated' by experts. These included the relative importance of freight as compared to the transportation of people, the weight of non-transport location factors as compared to transport cost, and share of non-tradeables per industry;

- parameters derived from recent Dutch bi-regional input-output tables (RUG/ CBS, 1999). These included the input cost shares for the 14 industrial sectors and the one household sector per region;
- econometrically estimated parameters. The latter include all substitution elasticities and the remaining parameters of the transport cost functions.

The last estimation was done for all 18 parameters simultaneously, minimising the squared *logsum* error of the actual and the predicted trade flows. The estimation uses the 588 observations on trade flows that could be derived from the 10 tables (namely three flows, on exports, imports and intra-regional deliveries, for 14 sectors in 14 regions). When all transport costs relate to freight, prices rise by 21 per cent after 100 km. When all transport relates to people, prices rise by 15 per cent after 100 minutes of travel. With an average elasticity of 12, total sectoral output halves after 21 km of freight transport and after 22 minutes of people transport.

After estimating all parameters, RAEM was calibrated on a projection of the Dutch economy for a reference scenario in 2020 (CPB, 1997), introducing an 'unexplained productivity' parameter per sector per region. Not surprisingly, these parameters were especially high for service industries, since SCGE models as yet are not capable of projecting structural shifts in the sector shares per region.

Besides, this first Dutch SCGE model had to interact with a labour migration model and an input-output expenditure model for labour migrants (Elhorst, *et al.*, 2000). Consequently, certain variables such as population and the number of firms had to be kept constant. This, unfortunately, meant that (internal) scale economies and (external) cluster economies could not be detected yet. Only the *short run* impacts discussed previously were estimated. Tables 5.3 and 5.4 show a summary of the outcomes.

Table 5.3 Change in Jobs per Region per Infrastructure Variant Compared to Reference Scenario, EC 2020

Δ Jobs	HIC	HHS	ZIC	ZHS	MZB	MZM
Northern Netherlands	650	1800	900	2000	3500	3100
Randstad	250	400	1200	1800	2200	2500
Flevoland	350	600	400	900	2100	2500
Rest of the country	-1250	-2900	-2500	-4700	-7800	-8100

First, Table 5.3 shows the spatial redistribution of jobs over the Netherlands due to executing each of the variants separately. Regions at the end of the line (the Randstad and the North) together with the region along the line (Flevoland) profit at the cost of the rest of the country. The rest of the country experiences a relative deterioration in its competitive position, especially on the economically largest market in the West of the country. Clearly the effect of speed is decisive, but also the trajectory of the variant plays a role. The Hanze-line variants HIC and HHS only increase the speed along an almost unchanged trajectory. Hence, especially the economic core area (the Randstad) hardly profits from this improvement, whereas it profits clearly more from the four other variants that involve new trajectories to the North.

The aggregate outcomes in Table 5.3 of course hide a substantial redistribution of jobs at lower spatial levels of analysis. The underlying material at the level of 14 sectors and 548 municipalities, for instance, shows that the bulk of the employment effect in the North relates to the services sector in the city of Groningen. This is not too surprising as only business travel times improve, while Groningen is by far the largest (service) city in the North and also enjoys the largest %-gain in travel time to the largest sub-market of the Netherlands, that is the Randstad. Other sub-regions in the North, such as those to the east and south of the city of Groningen show negative employment effects as their relative competitive position deteriorates compared to northern cities closer along the new infrastructure.

Besides these regional (re-distributive) effects, there are also important national (generative) effects, shown in Table 5.4. This is a little surprising, as economies of scale are not yet allowed. There is a minimal decrease in national employment (not shown), because labour becomes relatively more expensive compared to total intermediate inputs. Total output (GDP), however, increases as savings in transport cost lead to lower prices and more demand. Most of the lower prices are passed on to consumers who also enjoy their own direct transport cost savings, which result in an overall reduction of consumer prices (ΔCPI).

Besides lower prices, consumers and firms also enjoy a greater variety of available consumption and intermediate goods. For example, people along the Transrapid line will be able to go to the opera in Amsterdam and return the same evening, something that is hardly possible today. Because of the explicit utility function(s), RAEM, as any other SCGE model, is able to translate the utility gain in the equivalent consumer income increase (ΔY) that would have been necessary to reach the same change in utility (welfare). In the case of the Transrapid these gains amount to more than 250 million Euro yearly, which has to be compared with an investment cost of 5–7 billion Euro.

Table 5.4 Changes in Output, Prices and Consumer Welfare Per Variant Compared to the Reference Scenario, EC 2020

	HIC	HHS	ZIC	ZHS	MZB	MZM
ΔGDP (in%)	0.004	0.010	0.004	0.010	0.016	0.016
ΔCPI (in%)	–0.02	–0.06	–0.02	–0.05	–0.09	–0.09
Eqv. ΔY (in million euro)	64	156	56	153	262	251

Finally, there is the interesting phenomenon that the increase in output is much, much smaller than the decrease in the CPI. Part of this difference is explained by the peculiar implications of using iceberg type transport costs. Reducing iceberg type transport costs does not free up transport sector inputs, such as labour and intermediate inputs. Instead it results in less product being 'melted away' during transportation. This implies that the supplier needs to produce less to satisfy the same level of demand on the part of its customers. Hence, the consumption of non-transport products is able to increase more than the production of non-transport products. This unwarranted fact went unnoticed in the literature until now.

When a macro SCGE is used, this property does not pose a serious problem as the macro economic output is inclusive of transportation output that does (implicitly) reduce. In a multi-sectoral SCGE, however, these iceberg type transport costs imply a serious *mis-specification* as they lead to an underestimation of the output effects in the non-transportation sectors, especially in those sectors for which transport cost reduces most, whereas the opposite should be the case. RAEM, in fact, found the largest positive output effects in the public utility sector that only indirectly uses the transportation of people.

Some General Empirical Regularities

From the experience with RAEM and from earlier experience with potential models (Evers *et al.*, 1987) some general conclusions as regards the shape of the spatial pattern of the impacts of new infrastructure may be drawn. A fundamental difference appears between the impacts of point infrastructure (terminals and harbours) as opposed to the impacts of line infrastructure (roads, rail and waterways) (see Figure 5.2).

First, results are discussed when space is *isomorphic*, that is without differences in economic densities and without differences in transportation costs, which means that in equation (2) all Y_s are equal and all c_{rs} are strictly proportional to physical distance.

In the case of line infrastructure, Figure 5.2 shows the results for a simple uni-modal traffic plane, such as a rural region with only rural roads. In such an isomorphic plane, improvements in line infrastructure, such as a first stretch of a motorway, will produce *butterfly patterns* of diminishing impacts. In the smallest butterfly the impacts will be most positive. When the butterflies get larger the impacts will turn to negative and with still larger butterflies they will fade to zero.

In the case of improvements in point infrastructure, the isomorphic case is more complicated as terminals only have economic meaning in a multi-modal traffic case. Thus, the basic isomorphic point infrastructure case may best be imagined with two separate uni-modal isomorphic traffic planes that get a (better) connection in the point at hand. In that case, improvements in point infrastructure will produce simple *circles patterns* of diminishing impacts. As the circles become larger and larger, the impacts get smaller, become negative, and finally approach zero.

Second, Figure 5.2 shows results when (economic) space is *non-isomorphic*. The basic circular and butterfly pattern then get distorted. The way in which this happens is best understood with a close eye on (2).

In the presence of a major agglomeration, the areas at the opposite end of the infrastructure improvement profit most, whereas the areas close by will *not* profit despite this closeness. On the agglomeration side of the new line or new transhipment point, the pattern will be undistorted and show its basic circular or butterfly form. Comparable types of distortion result when barriers to spatial interaction, such as country borders, are included in the analysis.

Figure 5.2 is also useful when discussing infrastructure impacts at the level of a specific regional delimitation. Let us consider the outer circle or outer butterfly limits where the impacts reduce to zero. The area within this limit may be called the *influence area* of the infrastructure improvement studied. The total effect within this influence area may be either zero or positive. In the zero case, the infrastructure only has *re-distributive* effects within the influence are. In the positive case, the net positive effect is usually labelled a *generative* effect.

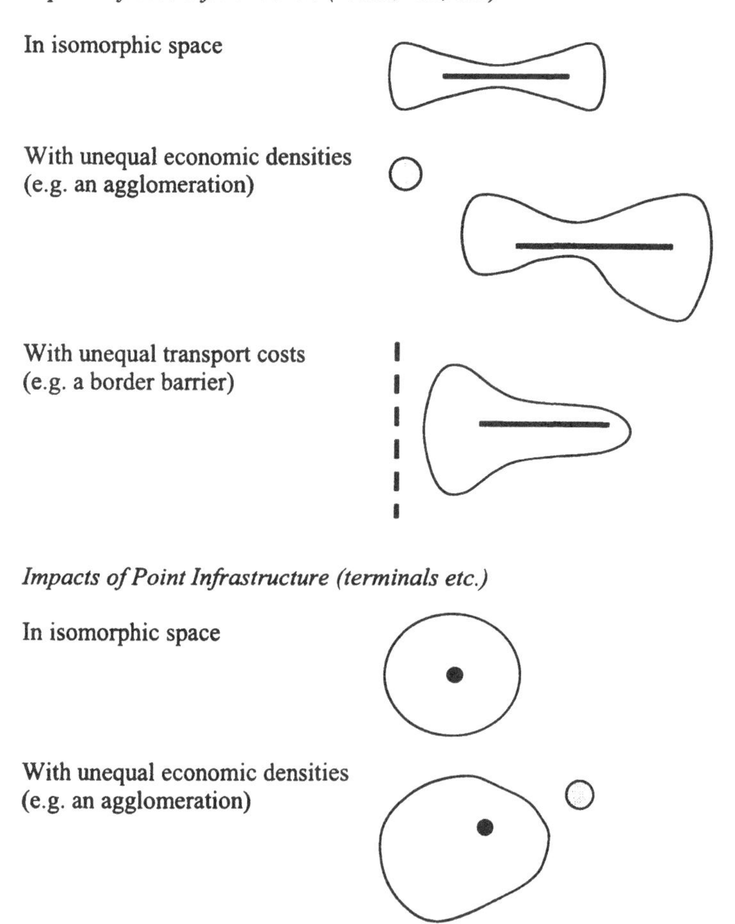

Figure 5.2 Different Spatial Impact Patterns of Transport Infrastructure

The literature appears to agree that the generative effects of new or improved infrastructure are minimal when mature, well-developed economies are involved (Blonk, 1979; Vickerman, 1991; Rietveld and Bruinsma, 1998). The re-distributive effects at lower spatial levels of analysis may be larger, especially when network bottlenecks are removed. The principle reason for this general conclusion lies in the small relative (%) reduction in generalised transport costs that is attainable in mature economies. One may only expect sizeable %-reductions when entirely new transport infrastructure with structurally lower transport costs or times, is under consideration. But even then, as in the case of the Transrapid rail variants, impacts may be moderate, as the %-reduction over the whole transportation chain is much lower than that over the new rail part of the chain. Moreover, mature economies typically supply competing modes of transportation (road next to rail) which further reduces the impacts of infrastructure improvements, especially when the latter concern the mode with the smaller modal share, as was the case with the Transrapid application.

The impacts of new infrastructure in developing countries – for the opposite reasons – often are considerable. Such countries have fewer or no competing transportation modes, whereas the existing modes are of such a low quality that most improvements tend to result in relatively large %-reductions in generalised transport cost.

Conclusion

In this paper we have given an overview of the different approaches found in the literature to estimate the economic impacts of investments in transport infra-structure. From this overview it can be concluded that land-use/transportation interaction (LUTI) models provide the best tested approach, which is most suited for infrastructure issues at the level of large urban conglomerations. Spatial computable general equilibrium (SCGE) models provide a theoretically more satisfying approach, which is especially suited to model the interregional impacts of new or improved transport infra-structure at a larger spatial scale.

This paper further discussed the recent Dutch construction and application of a 14–sector, 548–municipality SCGE model (RAEM). Both the construction and the application to a series of recent Dutch rail proposals showed this approach to be very promising, especially with regard to the possibility to produce a cost-benefit consistent valuation of consumer welfare changes. The application also showed that the standard SCGE use of iceberg type of transport costs leads to a serious mis-specification when multi-sectoral CGE models are being built.

Finally, the paper concluded with the circular and butterfly type of spatial patterns that will be found with Potential and SCGE type of models, and the paper presented an explanation for the small generative impacts as opposed to the sometimes quite large re-distributive impacts in the case of mature well-developed economies.

Note

1 The rail connection of this study includes the ZHS variant (see Table 5.2) analysed in the following section.

References

Aschauer, D.A. (1989), 'Is Public Expenditure Productive?' *Journal of Monetary Economics*, Vol. 23, pp. 177–200.

BCI/NEI (1997), *Ruimtelijke-economische verkenning van de Toekomstige Nederlandse Luchtvaart Infrastructuur*, Buck Consultants International/Nederlands Economisch Instituut, Nijmegen.

Berg, M.M. van den (1999), 'Location and International Trade, in Theory and Practice', Ph-D., University of Groningen.

Blonk, W.A.G. (ed.) (1979), *Transport and Regional Development*, Saxon House, Farnborough.

Bröcker, J. (1995), 'Chamberlinian Spatial Computable General Equilibrium Modelling: A Theoretical Framework', *Economic Systems Research*, Vol. 7, pp. 137–149.

Bröcker, J. (1998), 'Operational Spatial Computable General Equilibrium Modeling', *The Annals of Regional Science*, Vol. 32, pp. 367–387.

Bos, S. (1998), 'Direction Indirect, The Indirect Energy Requirements and Emissions from Freight Transport', Ph-D., University of Groningen.

CPB (1997), *Economie en fysieke omgeving, beleidsopgaven en oplossingsrichtingen 1995–2020*, Centraal Planbureau, Den Haag.

Dicken, P. (1986), *Global Shift: Industrial Change in a Turbulent World*, Harper and Row, New York.

DSC/ME&P (1998), *Review of Land Use/Transportation Interaction Models*, Report to SACTRA, Dept. of Transport, Environment and the Regions, UK.

Dixit, A.K. and Stiglitz, J.E. (1977), 'Monopolistic Competition and Optimum Product Diversity', *American Economic Review*, Vol. 67, pp. 297–308.

Eijgenraam, C.J.J. (1995), 'Macro-economische effecten van een infrastructuur project in Nederland', *Openbare Uitgaven*, Vol. 6, pp. 263–73.

Elhorst, J.P., Knaap, T., Oosterhaven, J., Romp, W., Stelder, T.M. and Gerritsen, E. (2000), *Ruimtelijk Economische Effecten van Zes Zuiderzeelijn Varianten*, REG-publicatie 22, Stichting Ruimtelijke Economie, University of Groningen.

Evers, G.H.M. and Oosterhaven, J. (1988), 'Transportation, Frontier Effects and Regional Development in the Common Market', *Papers of the Regional Science Association*, Vol. 64, pp. 37–51.

Evers, G.H.M., van der Meer, P.H., Oosterhaven, J. and Polak, J.B. (1987), 'Regional XII Impacts of New Infrastructure: A multi-sectoral potentials approach', *Transportation*, Vol. 14, pp. 113–26.

Ewijk, C. van, Hakfoort, J. and Schnieders, R. (2000), 'Infrastructuur en regionale groei; een empirisch onderzoek voor Nederland', 1973–1993, *Maandschrift Economie*, Vol. 66, pp. 357–83.

Fan, W., Treyz, F. and Treyz, G. (1998), 'Towards the Development of a Computable Geographic Equilibrium Model', Paper to the North American RSAI Meetings, Santa Fe.

FNEI (1984), *Economische betekenis van transportinfrastructuur*, Federatie van Noordelijke Economische Instituten, Groningen.

Fujita, M., Krugman, P.R. and Venables, A.J. (1999), *The Spatial Economy, Cities, Regions and International Trade*, MIT Press, Cambridge, MA.

Jones, S.R. (1981), 'Accessibility Measures: A Literature Review', *Transport and Road Research Laboratory*, Crowthorne.

Klundert, T. van de (1993), 'Crowding Out of Private and Public Capital Accumulation in an International Context', *Economic Modelling*, Vol. 10, pp. 273–284.

Knaap, T. (2000), 'Economic Geography and the Gravity Model', (mimeo) University of Groningen.

Knaap, T. and Oosterhaven, J. (2000), 'The Welfare Effects of New Infrastructure: An Economic Geography Approach to Evaluating a new Duch Railway link', Paper to the North American RSAI Meetings, Chicago.

Krugman, P. (1991), *Geography and Trade*, MIT Press and Leuven University Press, London.

Lee, S.Y., Haghani, A.E. and Byan, J.H. (1995), *Simultaneous Determination of Land Use and Travel Demand with Congestion: A System Dynamics Modeling Approach*, TRB Paper No. 950716, Washington D.C.

McCann, P. (1998), *The Economics of Industrial Location: A Logistics-Costs Approach*, Springer-Verlag, Heidelberg.

NEI (2000), *Vervoerswaarde Studie Zuiderzeelijn, Eindrapport*, NEI Transport i.s.m. Bouwdienst Rijkswaterstaat, Rotterdam.

NEI/TNO/RUG (1999), *Fundamenteel voorwaarts, Naar een praktisch werkbare en theoretisch gefundeerde benadering van voorwaartse economische effecten*, NEI Transport/TNO Inro/RUG FEW, Rotterdam.

Oosterhaven, J., Sturm, J.E. and Zwaneveld, P. (1998), *Naar een theoretische onderbouwde aanpak van voorwaartse economische effecten: Modelmatige definitie*, TNO Inro/ Rijksuniversiteit Groningen, Delft.

Pellenbarg, P.H. (1998), 'Het huidige belang van infrastructuur en vervoer voor regionale en nationale vestigingsbeslissingen', in J.P. Elhorst and Strijker D. (eds), *Het economisch belang van het vervoer, Verleden, heden en toekomst*, REG-publicatie 18, pp.51–65, Stichting Ruimtelijke Economie, University of Groningen.

Rietveld, P. (1989), 'Employment Effects of Changes in Transportation Infrastructure: Methological Aspects of the Gravity Model', *Papers of the Regional Science Association*, Vol. 66, pp.19–30.

Rietveld, P. and Bruinsma, F. (1998), *Is Transport Infrastructure Effective? Transport Infrastructure and Accessibility: Impacts on the Space Economy*, Springer Verlag, Berlin.

Rietveld, P. and Nijkamp, P. (2000), 'Transport Infrastructure and Regional Development', in Polak, J.B. and Heertje, A. (eds), *Analytical Transport Economics, An European Perspective*. Edward Elgar, UK.

Rothengatter, W. (2000), 'External Effects of Transport', in Polak, J.B. and Heertje, A. (eds), *Analytical Transport Economics, An European Perspective*, Edward Elgar, UK.

RUG/CBS (1999), *Regionale Samenhang in Nederland, Bi-regionale input-output tabellen en aanbod – en gebruiktabellen voor de 12 provincies en de twee mainport regio's*, REG-publicatie 20, Stichting Ruimtelijke Economie, University of Groningen.

Sturm, J.E. (1998), *Public Capital Expenditure in OECD Countries: The Causes and Impact of the Decline in Public Capital Spending*, Edward Elgar, Cheltenham, UK.

Toen-Gout, M.W. and Sinderen, J. van (1995), *The Impact of Investment in Infrastructure on Economic Growth*, Onderzoeksmemorandum 9503, OcfEB, Erasmus University Rotterdam.

Venables, A.J. and Gasiorek, M. (1996), 'Evaluating Regional Infrastructure: A Computable Equilibrium Approach', (mimeo) Report to the European Union.

Venables, A.J. and Gasiorek, M. (1998), 'The Welfare Implications of Transport Improvements in the Presence of Market Failure', (mimeo) Report to SACTRA.

Vickerman, R.W. (ed.) (1991), *Infrastructure and Regional Devel*opment, Pion, London.

Weber, A. (1909), *Über den Standort der Industrien*, Mohr, Tübingen.

Wilson A.G. (1998), 'Land-use/Transport Interaction Models, Past and Future', *Journal of Transport Economics and Policy*, Vol. 32, pp.3–26.

Chapter 6

The Economic Development Effects of Transport Investments

David Banister and Yossi Berechman

Introduction

This chapter examines the general question of whether, in advanced economies, transport infrastructure investments can engender economic development at the regional level, or merely facilitate its attainment when and if it transpires. It examines some key conceptual and analytical issues that underlie this question and, with the help of some empirical evidence, draws policy conclusions. Its purpose is to be thought provoking and challenging as new thinking is required to advance research on the links between transport investment and economic development (Banister and Berechman, 2000 and SACTRA, 1999). Conventional wisdom seems to be based too much on assumed relationships rather than clear logical thought.

For example, there is a prevalent belief among decision-makers and transport analysts that transport development plays a vital role in enhancing economic growth by lowering production and distribution costs, improving labour productivity, stimulating private investments and technological innovations. Underlying this conviction is the theory that the availability of fast, reliable and affordable transport historically has been the building block around which cities and regions have developed and flourished. The ability to move people and goods easily and economically is still used to explain the relative economic advantage of regions and states.

Although we challenge this widespread perception, it is important to note its policy implications. Proponents of this view tend to regard planned transport infrastructure investments as a key policy means for generating metropolitan, regional or national economic growth. Numerous statements by public officials and policy makers support this opinion and its corollary that the lack of transport investments will necessarily impede future growth and productivity improvements. By and large, these views are held as a 'truism' even though the available evidence on the subject is rather ambiguous. Moreover, in many cases these alleged 'economic development impacts' are used to rationalise capital investment projects, even when it is difficult to accept them on the basis of their transport effects. The EU has now realised the importance of this debate into the nature of the relationship between transport improvements and economic development (EU, 1997).

In the present context we define 'transport investment' as a capacity expansion or addition to an existing network of roads, rail, waterways, hub terminals, tunnels, bridges, airports and harbours. Transport capital improvements are carried out incrementally, project-by-project over many years, and each new facility constitutes but a segment of a larger network. Hence, while each new project needs to meet evaluation criteria, its primary transport impacts are appraised relative to the in-place network in terms of improved travel times, costs and traffic volumes over the network. 'Economic growth' is defined as a continuous process of annual increases in per capita income, factor productivity, national, state or regional product and employment. Mainly for practical reasons, employment is the most commonly used measure of growth[1] in empirical studies. In contrast, the concept of 'economic development' is used primarily when examining the effect of additional investment in specific types of infrastructure on the urban and regional economy. This concept also encompasses some non-growth objectives, such as changes in urban form, equity effects and the reductions in environmental quality. Thus, changes in regional employment (by type), adjusted for changes in location of firms and households, are used as a proper measure for assessing the effects of transport infrastructure investment on local economic development. In general, we regard the change in economic opportunity resulting from accessibility improvements, which is capitalised in the form of greater use of input factors, expanded output or enhanced welfare, as economic development.

Given these definitions, is there a sound rationale for the above transport-economic development contentions? If so, what is the underlying mechanism that links new transport infrastructure investments with economic development? How does this linkage manifest itself in the face of emerging forms of regional and national economies? If it exists, how can we model and measure it? What does empirical evidence tell us about the real-world impact of transport investments on development? Given these questions we also need to ask what are the implications if these alleged linkages are rather loose or insignificant? Will it then require a reassessment of the ways transport projects are evaluated? We will examine these questions only very succinctly, referring the reader to the much greater detail in the book by Banister and Berechman (2000).

The Causality Question

The standard approach to the linkage between transport investment and economic development suggests that transport investments generate two major effects: 'indirect effects', mainly economic multiplier and environmental impacts, and 'direct effects' defined in terms of accessibility improvement impacts. The multiplier effect results from the public-work nature of the investment as it generates employment and income in the local area, and these benefits last throughout the project's implementation period. It also translates into longer-term employment at the investment (e.g. airport), with further impacts on indirect investment related activities and the wider economy (induced employment). The second category contains the transport benefits whose magnitude and spatial distribution depend on the specific transport facility (e.g., rail, port, or highway), the network and various regional features. These benefits, in turn, are assumed to generate long-term development effects as they improve the economic performance of individuals and firms and generate more efficient location patterns. Figure 6.1 presents these relationships.

We challenge the view depicted by Figure 6.1 of the economic development outcomes from transport investment. As economic growth is a long-term phenomenon, to regard short-term multiplier effects as contributing to economic growth negates its basic definition, and the longer-term effects are less clear in both their scale and impact. In addition, these multipliers are often taken from previous studies, do not relate to the specific project under investigation, and again are taken on 'trust' rather than questioned or debated. Moreover, these multipliers ignore the specific location and network impacts of transport improvements and their unique effect on the economic behaviour of firms and households. In general, various forms of government sponsored work programmes, unrelated to transport, can generate short-term income, employment and other local economic effects, but the longer-term or wider impacts on the local and regional economy are less clear.

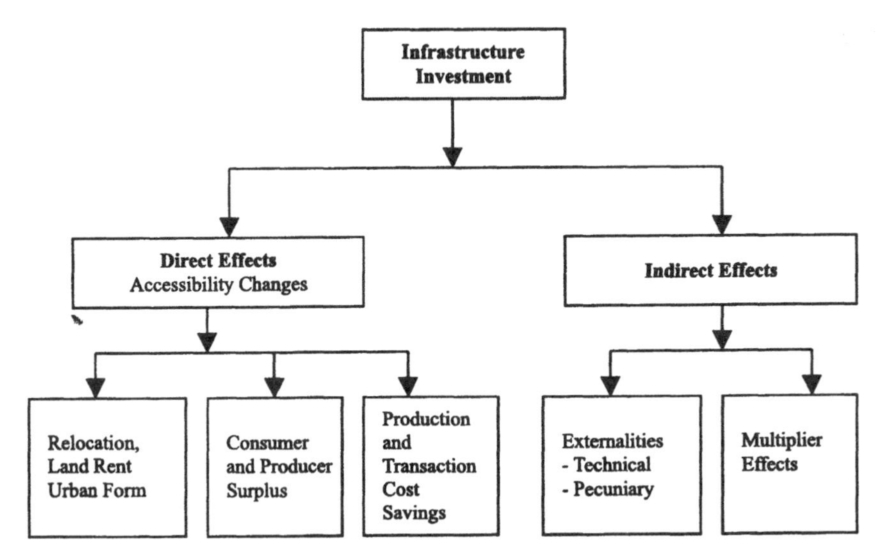

Figure 6.1 Conventional View of the Effects of Transport Infrastructure Investment

Turning to the direct effects, we challenge the assumption implicit in Figure 6.1 that accessibility improvements from transport capital investment necessarily promote economic development. This argument can be presented in two steps. First, there is the declining role of accessibility improvements in the contemporary economies of cities and regions. In already well-connected networks, an additional link or increase in capacity will have a marginal effect on overall levels of accessibility in most cases. There will of course be particular situations where a single improvement will have a substantial impact (e.g. the Channel Tunnel or the Øresund Link). This means that accessibility improvements must be supported by other prevailing positive market conditions in order for these combined benefits to potentially generate economic development impacts.

Because of key demographic, transport and economic trends, contemporary Western economies are much less impacted by transport improvements than even a few decades ago. There is a marked change in the relative importance of work related trips, which traditionally were the major cause of congestion, thus of capacity improvements. Data from the UK show that of the major categories of daily trip purposes in 1997/99, work and business travel accounts for 22 per cent of all trips, down from 26 per cent fifteen years earlier. A similar phenomenon has been observed in the USA (DETR, 2000).

This trend is strongly related to another major change, namely the development of highly dispersed employment patterns, mainly at the expense of the central business district (CBD). As a result, commuting patterns have become highly complex, with cross commuting becoming more important than travel to city centres. Findings on auto commuting trip times for the twenty largest USA metropolitan areas show that average trip times have remained constant during the eighties and nineties or were reduced (Gordon and Richardson, 1994). Apparently, the market operates through the relocation of firms and households to achieve the balance of keeping commuting times within tolerable limits.

The restructuring of the economy in the post-industrial society is another major change. In today's economy, the main source of profits and market dominance is knowledge and information, a major part of which is unrelated to transport. New information and telecommunication technologies are considered more vital to improved production and distribution processes than transport (Graham and Marvin, 1996).

The last structural change discussed here relates to the effect of transport on the environment. In 1996, transport was responsible for over 25 per cent of world primary energy use and 23 per cent of CO_2 emissions from fossil fuel use, and developed countries contribute the majority of this figure. Recently, environmental arguments have been linked to those of sustainability, connecting environmental concerns with those of economic development and equity. To achieve the objectives of sustainable development, individuals and firms must carry out their daily activities differently, mainly in ways that significantly reduce travel (Banister *et al.*, 2000).

In general, a large proportion of all transport investments is made to improve accessibility and alleviate congestion, mainly for daily commuting and freight movement. The simultaneous operation of the above trends lessens the alleged linkage between accessibility improvements and economic development as described in Figure 6.1. How then can transport investments induce economic development? To answer this question we now introduce an alternative analytical framework, depicted in Figure 6.2.

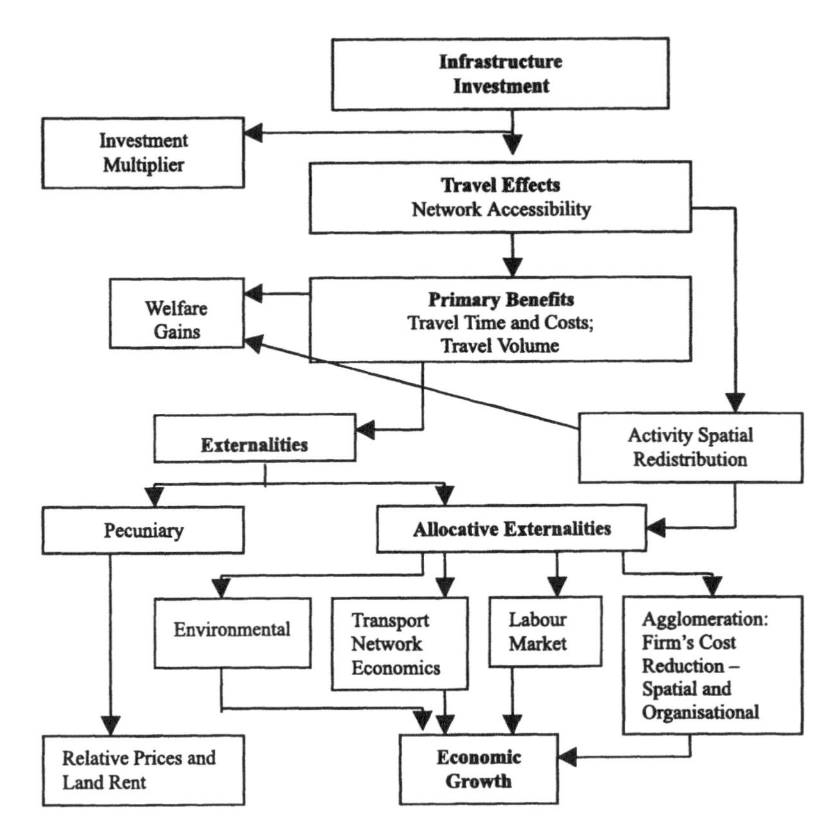

Figure 6.2 Proposed Linkages between Transport Investment and Economic Development

Figure 6.2 highlights the idea that the main output from a transport investment, is *network accessibility* improvement. Assuming a positive net value of these effects they also represent welfare gains to households and firms. Subsequently, two additional effects may arise. First is the impact of network accessibility improvements on activity location, which if it ensues, may improve spatial patterns and economic efficiency. Later, we return to this issue. The second potential result is economic development. This effect is predicated on the presence of certain market conditions, labelled in Figure 6.2 as 'allocative externalities'. These effects emanate from the non-compensatory action of one economic entity on the utility level of another, which in turn, can affect the efficient allocation of resources in the economy. Traffic congestion is an example of negative allocative externalities, whereas firms' agglomeration represents positive ones.

Transport accessibility improvements can potentially trigger several major positive externalities, which are susceptible to accessibility enhancement. In turn, they can boost productivity, reduce production costs and promote more efficient use of resources. Collectively, these changes can bring about economic development as defined at the outset. And these benefits must be *in addition* to the primary accessibility improvement benefits and not merely their market capitalisation.

To summarise, the main argument regarding economic development ensuing from transport infrastructure development is that the mechanism that transforms accessibility benefits into economic development benefits is the presence of positive allocative externalities in specific markets, which are amenable to improved accessibility. The scale, spatial and temporal distribution of these externalities will affect the magnitude and scope of economic development, given the transport investment. Noticeable examples are labour market economies; an economy of industrial agglomeration; and transport market economies. An important example of the latter is when two disjointed networks are linked by a newly constructed facility, thereby opening up for trade previously non-trading markets.[2] Another example is when a new freight terminal enables intermodality (say, between truck and rail), which improves 'just in time production', thereby reducing inventory costs to producers.

Is Transport Investment a Generator or Supporter of Economic Development?

The conventional view regards transport investment as a *sufficient condition* for economic development in the sense that its development generates growth. At times, transport development is also regarded as a *necessary condition* in the sense that unless further transport investments are made, development will be retarded. Are these claims true or do they hold only in some very special cases? The principal argument made here is that transport investment is neither a necessary nor a sufficient condition for economic development on its own. As depicted by Figure 6.2, it is only when positive externalities can be established that transport investment can potentially generate growth.

In general, transport infrastructure acts as a *binding constraint* on the local economy. That is, if transport capacity is rather limited, given the level of economic activity in an area, its further expansion will enable the region to become more competitive relative to adjacent regions. Can this improved competitiveness be regarded as a long-run economic development? In an open economy, the answer must be negative as factor mobility, mainly of capital and labour, and the frequently used economic preferential policies (like tax incentives), can perhaps make such an edge significant, but only in the short-run. On the other hand, to regard this constraining characteristic of the transport system as unimportant or inconsequential is a serious mistake. In many regions (e.g. New York Metropolitan Area), regional competitiveness can be substantially enhanced if critical transport investments (in particular, those related to freight movement) were implemented. Given budget constraints, as long as the time and costs savings from transport investments yield a positive social rate of return, they should be undertaken regardless of whether economic development follows.

Can we regard transport investment as an economic generator in cases where positive market externalities can be ascertained? The answer is a 'qualified yes', since we need to recognise that the amount of growth is a function of the degree of improved accessibility produced by a project (see Figure 6.2). In well-developed economies, where the in-place transport network is quite extensive, additional accessibility, even from a large-scale capital project, cannot be significant. Additionally, even this economic development is confined primarily to transport intensive sectors where accessibility matters (e.g. retail, agriculture or food distribution). Since contemporary economies are fuelled primarily by innovations in the information and data processing sectors overall, the amount of transport generated growth is rather limited (Swyngedouw, 1993).

All in all, there is plenty of evidence showing long-term regional economic development despite the lack of any significant transport investment. Apparently, many other forces, ranging from a sharp rise in international trade, technological innovations, human capital betterment to successful local economic policies, can all support growth. Hence, in general transport development serves as a growth supporter and *not* as a growth generator. By and large, within a region with a reasonable level of transport accessibility, development is achieved by an assortment of forces and policies, not necessarily transport related.

Some Conceptual and Empirical Disputes

Given the conventional and alternative views on the relationship between transport development and economic growth, we now discuss five further misconceptions that frequently appear in the popular and professional literature.

Accessibility benefits and potential development benefits can be added to produce total benefits from a transport capital investment

It has been argued that in the evaluation of transport capital projects, only direct travel time and costs savings should be regarded as benefits from the project, since all other alleged benefits, result from the capitalisation of these costs savings (Mohring, 1993). The inclusion of other effects, like potential economic development, as additional benefits, amounts to *double counting* of benefits. In the absence of positive externalities, the welfare gains (actually the change in consumer surplus) resulting from the primary transport benefits represent *total benefits* from this project (see Figure 6.2).

Transport improvements are associated with long-term economic development; hence they must also generate it

A number of authors have found correspondence between the timing of major transport improvements and long-term cycles of economic development. They have deduced, therefore, that the former is a major cause of the latter (Garrison and Souleyrette, 1996). But what is the cause and what is the effect? Can key transport advances underlie economic development (the 'generator' function) or a surge in interregional and international trade, which strains the capacity of the in-place transport systems, and this, in turn, leads to the requirement for new transport improvements and innovations to sustain this development (the 'supporter' function)? The empirical historical research literature on this question is quite equivocal. To illustrate, in a renowned study, Fogel (1964) has analysed the impact of railroad development on American economic growth during the 19th century. He concluded that while railways had a primary impact on the costs of transport and that social savings have resulted from the movement of agricultural output by rail, 'no single innovation was vital for economic growth during the 19th century'. Economic growth was primarily a consequence of the knowledge acquired in the course of the scientific revolution and this was the basis for a multiplicity of innovations. Thus, rail development in the U.S. has helped shaping growth in a particular direction but was *not* a prerequisite for it.

Transport improvements produce efficient spatial patterns thereby generate economic development

A conventional view states that improved transport will stimulate efficient spatial patterns of households and businesses, which, in turn, will spur economic development. This inference is supported by a vast amount of theoretical analysis, the main conclusion of which is that improved accessibility will encourage further activity decentralisation and, at the same time, will intensify agglomeration and urbanisation economies (Anas *et al.*, 1998). How valid is this view in light of available empirical evidence on metropolitan expansion? Gordon and Richardson (1994), for example, have found that in the Los Angeles metropolitan area over the last two decades, average travel times and congestion levels have declined even though no significant investments in transport infrastructure facilities were made. The main reasons are the decentralisation of firms and the formation of new sub-centres of business and employment. By and large, other factors, such as the availability of skilled labour, override high transport costs. It appears that, in the long run, the cumulative accessibility improvement impacts from transport development tend to encourage more dispersed spatial patterns that may not be regarded as efficient. In general, available empirical evidence casts a strong doubt on the effectiveness of *specific* transport projects to influence relocation decisions and land use changes (Giuliano, 1995).

Transport's share in GDP can be used as an economic growth indicator

In the U.S., the percent of transport costs of total production costs by industry has been computed to be 313 billion U.S. Dollars, which is about five per cent of the USA GDP (1992 data). This percent share is often equated with the concept of value-added, to explain the economic value of transport (Bureau of Transportation Statistics, 1998). This approach is erroneous since value-added is a measure of the contribution of an industry to the economy over and above the value of the input factors it uses for production. The cost of use of a particular input factor can comprise a large share of GDP while its value-added can be rather low.

Transport share in industry's costs can be used to measure the economic contribution of transport

A common practice is to calculate the cost of transport as a percent of one U.S. Dollar of output of an industry and subsequently to argue that a one U.S. Dollar increase in this industry's output will require an equivalent per cent increase in transport services (Bureau of Transportation Statistics, 1998). For this claim to be correct firms' cost functions must be assumed completely elastic (having the property of constant returns to scale), thereby ignoring the prevalent cases of costs savings due to scale effects and factor substitution. A related mistake is to argue that a given reduction in transport costs will bring about an exact increase in output or productivity.

Empirical Evidence

Attempts have been made in recent years to validate and measure the impact of specific transport investments on economic development mainly at the local and state levels. In this section we very briefly review three studies of rail, highway and airport development.

The Buffalo Light Rapid Rail Transit (LRRT) is a massive capital investment project (about 520 million U.S. Dollars, in 1982 prices) whose explicit objective was to revitalise the Buffalo CBD by offering high quality rail accessibility, thereby encouraging economic development. Berechman and Paaswell (1983) have studied this project and have derived two main lessons. First, that improved rail transit accessibility in a region with high quality highway access, is neither a necessary nor a sufficient condition for revitalising the CBD. Second, the lack of regional or citywide coordination of policies to ensure the attainment of the LRRT objectives was probably the most serious threat to the revitalisation of the Buffalo downtown. Conflicting highway, parking, transit and zoning policies are examples of this problem.[3]

The Orbital Motorway around Amsterdam was completed in 1990. This highway is only 5 km from the city centre and diverts through traffic from the city streets. It was suggested that this project would have a major impact on office location and commercial development in the region. But a study by Bruinsma *et al.* (1996) has found no significant impact on office rents and locations. In fact, other factors (like building amenities) had a greater impact on office location decisions than improved accessibility.

The Proposed Terminal 5 at London Heathrow Airport (LHR-T5) is an example of a major airport development. The growth and restructuring of domestic, European and intercontinental aviation markets have created a business opportunity for development at hub airports such as Heathrow.[4] Thus, Heathrow's management believes that it would be commercially advantageous to expand the capacity of Heathrow by constructing a new terminal alongside the four existing ones. A key issue in the decision-making process is the role of the airport as a generator of economic activity, mainly in the airport's region. Using employment as a key development measure, studies by Pieda (1994) have raised some difficult questions on the airport's ability to boost employment. Most important is the fact that a sizeable increase in total employment attributed to the investment is a direct function of the expected increase in air traffic at Heathrow. This means that the real generator of economic development, in the form of induced employment, is an external factor (i.e. the expected growth in aviation traffic) and *not* Terminal 5 investment per se. Thus, this project is neither a necessary nor a sufficient condition for economic development. Rather, the airport's current capacity represents a *constraint* on its ability to accommodate the expected increase in traffic at a desired level of service. It is important to point out that without Terminal 5, traffic may divert to the two other major airports in the region (Stansted and Luton) which presently are under utilised, thereby improving air and ground traffic distribution in the London and South East Region.

Transport Investment and Economic Development: The Decisive Role of Policy Design and Evaluation

Two major concerns have arisen from this critique. The first concerns the regional dimension and the necessary spatial conditions for economic development, in the context of the creative learning capacity. This is of particular importance within the context of the EU strategy on the development of the transport TENs. The second presents a new proposal for project appraisal that explicitly includes the necessary economic development conditions. In both cases, they represent clear developments in our understanding of the links between transport investment and economic development, given the rejection of the conventional arguments.

The Regional Dimension

It has been argued that changes in accessibility resulting from transport infrastructure investment cause a redistribution of employment between regions. It is unclear whether the changes in accessibility also create new activities, which we would call economic development (see Figure 6.3, Box 2+3). The conclusions reached here suggest that at the regional level redistribution will take place, often to the further advantage of the already accessible core parts of the country. Physical accessibility changes operate in two directions, both in and out of the region, and the analysis should identify the net effects (Oosterhaven and Knaap, 2002).

But, transport accessibility must also be seen as part of a much wider concept of accessibility that includes availability of skilled labour, good quality locations, entrepreneurial skills, the availability of capital, and the necessary supporting infrastructure. It seems that the actors need to be able to manage the changing internal and external relationships to achieve the best economic output – this is what some (e.g. Storper, 1993) have called the creative learning capacity. Changes in transport accessibility resulting from infrastructure investment forms just one element in that process.

Open Dynamic Systems	
1.	**2.**
Economic self-sufficiency in these regions with production under strong local ecological controls. *Transport investment important and will have the maximum impact in these systems.*	International and national markets with strong conditions for further growth. *Transport investment already at a high level and will support development– but not a necessary condition.*
Inaccessibility	**Accessibility**
Poor transport may have contributed to decline, but on its own it will not revive these regions. *Isolated static local areas with declining economies.*	*Infrastructure good along corridors, but further investment will have little impact, as economic conditions are weak.* *Accessibility restricted to corridors – regions in decline, except at key interchanges.*
4.	**3.**
Closed Static Systems	

Figure 6.3 Transport and Economic Development at the Regional Level

In Figure 6.3, movement along the accessibility axis will not achieve economic development on its own. It is only where there are open dynamic systems (the economic and political sets of conditions are present) that it has a real impact, particularly where the existing infrastructure provides only poor levels of accessibility (Quadrant 1). In dynamic systems with high levels of accessibility it will support growth (Quadrant 2), but in the lower part of the figure (Quadrants 3 and 4) improving transport accessibility is not a sufficient condition for economic development (for the lack of the economic, investment and political sets of conditions).

The capacity of the system has to be enhanced so that it can respond – i.e. move to the top part of the figure. This means that the skills and knowledge base has to be raised (i.e., introducing economic conditions). Strong links with local research and the informal contact networks have to be established, and dynamic transactional relationships between manufacturers, local suppliers and customers are required. These are the key elements of the creative learning capacity that is a necessary prerequisite for economic development at the regional level (Figure 6.3). Quite often, various stakeholders have conflicting interests and agendas, thus making the design of coordinated and harmonious policies rather difficult. Even if underlying economic conditions are favourable, political conditions will ultimately determine the degree to which economic development outcomes are attained.

At least three types of policies are vital for the attainment of potential development benefits. The first is investment policy, which determines attributes such as mode type, the investment's scale, facility location and function in the larger network. Second, there is the regional economic policy, which influences firms' location, labour market conditions and land market economies. Lastly, there is the general public policy-making whose essential task is to resolve conflicts among stakeholders. In Western democracies, the ability of governments to coordinate their activities and design complementary policies to gain maximum economic development effects from their capital investments is fairly limited. Paradoxically, however, it is only in democratic societies that economic development reaches its maximum potential.

A New Proposal for Project Appraisal

A major criticism of the macro level analysis is the conclusion that finding a rate of GDP growth from a certain increase of total capital accumulation actually tells us little about the contribution of the next infrastructure project. That is, some projects have a high and positive impact on GDP growth whereas others may not. All that the macro level analysis says is about the *average* contribution of capital accumulation. But since infrastructure investments are carried out on an individual project basis, we need to analyse each separately where the general rate of return (or rate of GDP growth) from total public capital formation may at best serve as a kind of a guideline.

A further issue to contemplate is that analysis of all planned infrastructure investment projects is done *ex-ante* with little certain information about future trends and development. Hence there is a need to consider specific issues that may affect the potential impact of this investment on economic development. First, there is the need to consider each transport project within the framework of a local, regional or even national network. For development to occur, it must be the result of improvements at the network level and not at the single project level. The effect of a given investment on development can be dramatically different when, for example, it links two disjoint networks rather than being only an additional link in an established network. This was a clear conclusion from the EU COST328 Action on Integrated Strategic Infrastructure Networks in Europe, where the value added from individual links to the network formed the basis for developing evaluation processes that linked projects to programmes and policies, as well as determining critical success factors and decision making processes (EU COST328, 1998).

Prioritising objectives and criteria for project appraisal constitutes a second set of issues that decision makers need to be very clear about. We would propose a three-step procedure:

- the majority of benefits need to be transport related, since otherwise why invest in transport facilities in the first place? Cost benefit analysis should remain the key method for transport evaluation;
- the need to avoid double counting in measuring non-transport benefits must be clearly recognised, and explicit measures taken to highlight situations where this occurs (see earlier discussion);
- the need to show a functional linkage between primary transport benefits (e.g. accessibility improvements) and potential economic development effects should be demonstrated on a project by project basis.

If transport investment is to take place, we would suggest a twin approach where conventional cost benefit analysis is carried out on the project to determine the user benefits and costs of investment. To achieve a given rate of return, this analysis may account for some or all the necessary returns. If there is a shortfall, then a complementary analysis needs to take place that takes a wider view of the investment proposal. It could be argued that this complementary analysis should become an integral part of all evaluation, not just where the transport analysis fails to meet agreed criteria.

As itemised in Figure 6.4, this would include the contribution of the project to the transport network as a whole through network analysis. This is not difficult to carry out, either through existing land use and transport models or through network based general equilibrium models (Oosterhaven and Knaap, 2002). The key issue here is to identify the value added, not just in terms of physical factors (e.g. accessibility and time savings), but also in terms of the role of key actors in taking advantage of the new network integration. Network integration is demand-led within a market environment. Although the actors can facilitate integration through regulation, price, location and other complementary policies, it is the user of the network that primarily determines the level of integration. The freight sector best illustrates this point through its reorganisation. Value added is demonstrated in the form of the new flexible production processes with outsourcing and decentralisation, together with new management structures. It seems likely that other sectors (e.g. passenger transport) will adapt in the same way so that the integrated services will respond to the demand of users for high quality 'seamless' travel (e.g. in the leisure sector). This is the customer driven network.

Full network integration requires a linking of transport networks, together with economic, cultural and other networks. All of these networks interrelate, and it is difficult to apply one form of evaluation. Even if it was possible to develop a unified evaluation tool for network integration, the product is likely to be technocratic and only able to tackle part of the problem. This is a feature of current methods that mainly address a single mode in the context of a single project, concentrating on only a limited number of impacts (e.g. the physical infrastructure). We would propose a multiplicity of approaches and methods, and the analysis should be carried out on the functioning of networks in particular contexts (Banister *et al.*, 1999).

In addition, the value added of the project would be assessed through its contribution to local employment, the potential for increases in productivity, and the environmental impacts. To some extent, these issues have been addressed, but it is here that more sophisticated multiplier analysis could be carried out, in particular exploring the relationships between transport investment, output, productivity, and employment. It is unclear whether benefits from transport investments are capitalised through lower prices, through higher levels of employment, through higher wages, or through increased profits, or through different combinations of these possibilities.

Finally, the evaluation would also investigate the distributional impacts in terms of the spatial effects on the regional and local distribution of services and facilities, and the social impacts. As with all decisions, there are important spatial and social impacts. So the appraisal would need to address the distributional effects in a quantitative and a qualitative framework. Some effects are not difficult to identify in terms of the spatial impacts, as the core areas become more accessible, often at the expense of the peripheral areas, and as the buoyant local economies increase their dominance. Other impacts, particularly those relating to low income groups or 'disadvantaged' groups are much less easy to identify. Hence, in much of the regeneration literature, investments are made on the basis that the local population (often disadvantaged) will benefit. In reality, it is outsiders that benefit, and this in turn leads to greater social polarisation (Massey and Swyngedouw, 1993).

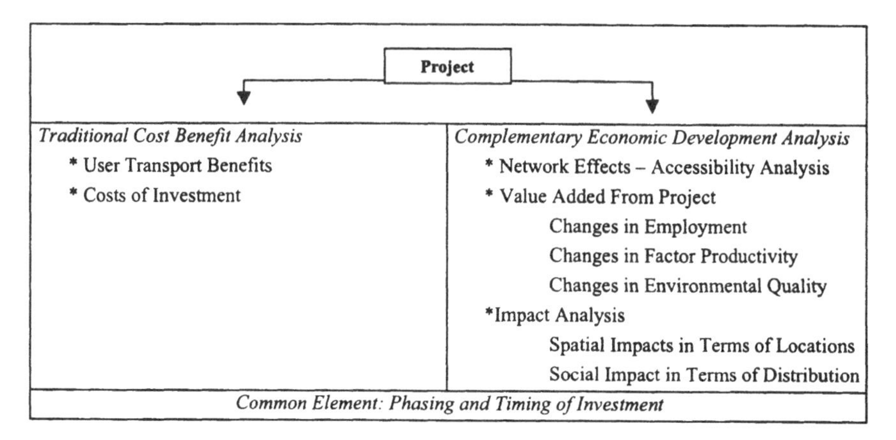

Figure 6.4 A Suggested Twin Approach to Project Appraisal

The contribution that the components outlined in Figure 6.4 make to the overall project assessment all need to be measured. If additional benefits can be shown and are sufficient to raise the benefit-cost analysis above the crucial level in terms of the rates of return, then investment could take place. This more complex type of analysis seems to be increasingly important, as the conventional benefits from any proposed transport investment may be providing an ever-decreasing proportion of the total returns, particularly where there is already a high quality transport network available.

Vickerman (2001) takes a broader interpretation of economic development to include all public expenditure on infrastructure, and he develops a three stage conceptual model. First, the objective of the transport intervention is identified, and secondly the spatial impact of the project is established. Here, he argues for all potentially affected regions and areas to be included as much analysis concentrates only on the impacts in the immediate vicinity of the investment. Thirdly, the sectoral impact should be assessed, as this reflects the traffic mix, the purpose of travel, the affected industries, and the cost structures of those industries in terms of their transport costs relative to the value-added and the price/cost margins in the sector. The aim here is to ensure that all the wider impacts are measured, not just the transport impacts. Even then, there are many potential economic development impacts that are ignored (see Figure 6.2), but the strength of the conceptual model is that socially optimal pricing could be used as a means to increase efficiency and encourage economic development rather than relying solely on infrastructure investment. It might also be a much cheaper option as it would tend to reduce barriers behind which inefficiency and imperfect competition takes place.

Conclusions

Do transport infrastructure investments generate national and regional economic development? This question remains an important policy issue as planners and policy-makers repeatedly provide a strong affirmative answer that is based on popular writings and basic intuition. But reality is not that simple, as the empirical evidence on the causal link between transport investment and development, at best, is equivocal. What then are the contributions made in this chapter?

- while at the national or state level capital stock expansion seems to be correlated with economic growth, transport development is carried out incrementally, by the implementation of individual projects. Hence, it is at the project level that a link must be established between the project's primary accessibility benefits and economic development. It is for this reason that a careful microeconomic analysis must be carried out to ascertain causality;
- transport capital projects can be justified only if they generate sufficient transport benefits. If not, attempts to rationalise their implementation on the basis of alleged economic development benefits is fundamentally wrong and is likely to lead to the implementation of inferior transport projects. Put alternatively, transport investments should be carried out, first and foremost, on the basis of the social rate of return from their primary transport benefits. Expected economic development effects are a secondary decision criterion that cannot replace transport output evaluation;
- by and large, transport capacity investments serve as supporters of regional economic development which, almost universally, is spurred by non-transport factors. It is only when certain conditions, related to the impact of accessibility benefits on market externalities, are met that transport development can potentially generate economic development. If such conditions cannot be shown, the frequently used practice of adding up accessibility benefits with other non-transport benefits amounts to double counting;
- even when positive externalities can be ascertained, critical policy conditions must prevail in order for development benefits to materialise. Even then, they may not be guaranteed. This is particularly important at this time and age when major policy and logistic efforts are made to decouple economic growth from further increases in passenger traffic and freight movement;
- to provide a solid foundation for the potential relationship between transport investment and economic development and as an objective for future research, more *ex post* type studies of transport investment projects must be made. Presently, our understanding of *actual* economic development benefits from transport improvement projects, are limited and largely unsubstantiated. Moreover, we understand only poorly the geographical incidence of economic development benefits. Are they confined to areas adjacent to where

the transport project takes place or do benefits also spill over to distant regions? More generally and within the EU context, does the centre gain at the expense of the periphery or does the opposite occur? There is an urgent need to monitor differences between expectations from transport investments and actual outcomes.

The principal message from this chapter is that it is *not* that transport projects are unable to bolster economic development benefits or enhance productivity and regional competitiveness. Rather, it is a matter of identifying and substantiating the presence of some necessary market conditions. In their absence, transport investments at best would generate only accessibility and other transport benefits. On the other hand, if these forces are indeed present and if suitable reinforcing policies are designed and implemented, a transport project can potentially promote local and regional economic development, and perhaps also contribute to an improvement in environmental conditions.

Notes

1 The use of employment as the key growth indicator is also politically motivated. Politicians are often sensitive to the level of unemployment among their constituents and frequently endorse or reject a transport investment project on the basis of its alleged capability to promote employment within their region.
2 An interesting example here is the Øresund Link. This is a transport corridor connecting Gothenburg and Stockholm with Malmo and Copenhagen by means of a four-lane highway and a dual track railway. It includes a 7.8 km bridge and a 3.4 km tunnel between Denmark and Sweden and has been open for traffic since June 2000.
3 Presently a consultant, hired by the controlling agency (NFTA), has recommended extension of the system in the metro area by 29 miles to boost its cost effectiveness. The alternative proposal is to completely shut the system down for lack of operational and maintenance funds (Buffalo News, July 17, 2000).
4 The Inspector's report was submitted to the Secretary of State for Transport (2000), and his decision was made in 2001 when the new terminal was approved. Construction will start in 2002.

References

Anas, A., Arnott, R. and Small, K. (1998), 'Urban Spatial Structure', *Journal of Economic Literature*, Vol. 36, pp.1426–1464.

Banister D. and Berechman J. (2000), *Transport Investment and Economic Development*, University College, London Press, London.

Banister, D., Maggi, R., Nijkamp, P. and Vickerman, R. (1999), 'Actors and Factors in the Integration of the Strategic Infrastructure Network in Europe', in Beuthe, M. and Nijkamp, P. (eds), *New Contributions to Transportation Analysis in Europe*, Avebury, Aldershot, pp.17–44.

Banister, D., Stead, D., Steen, P., Åkerman, J., Dreborg, K., Nijkamp, P. and Schleicher-Tappeser, R. (2000), *European Transport Policy and Sustainable Mobility*, Spon, London.

Berechman, J. and Paaswell, R, (1983), 'Rail Rapid Transit Investment and CBD Revitalization: Methodology and Results', *Urban Studies*, Vol. 20, Nr. 4, pp.471–486.

Bruisma, F., Pepping, G. and Rietveld, P. (1996), 'Infrastructure and Urban Development: The Case of the Amsterdam Orbital Motorway', in Batten, D. and Karlsson, C. (eds), *Infrastructure and the Complexity of Economic Development*, Springer Verlag, Berlin, pp.231–249.

Bureau of Transportation Statistics (1998), *Transportation Statistics Annual Report 1998*, US Department of Transportation, Washington DC.

DETR (2000), *Transport Statistics Great Britain 2000*, Department of the Environment, Transport and the Regions, The Stationery Office, London.

European Union (1997), *The Likely Macroeconomic and Employment Impacts of Investments in Trans-European Transport Networks*, Commission Staff Working Chapter SEC(97) Vol. 10.

European Union COST (1998), *Integrated Strategic Infrastructure Networks in Europe*, Final Report of the COST328 Action, Luxembourg.

Fogel, R.W. (1964), *Railroads and American Economic Growth: Essays in Econometric History*, John Hopkins, Baltimore.

Garrison, W. and Souleyrette R. (1996), 'Transport Innovation and Development: The Companion Innovation Hypothesis', *Logistic and Transport Review*, Vol. 32, Nr. 1, pp.5–37.

Giuliano, G. (1995), 'Land use Impacts of Transport Investments: Highway and Transit', in Hanson S. (ed.), *The Geography of Urban Transport*, Second edition, Guilford Press, New York, pp.305–341.

Gordon, P. and Richardson, H. (1994), 'Congestion Trends in Metropolitan Areas', in *Curbing Gridlock, Peak Period Fees to Relieve Traffic Congestion*, Vol. 2, Transport Research Board, Special Report 242, pp.1–31.

Graham, S. and Marvin, S. (1996), *Telecommunications and the City: Electronic Spaces, Urban Places*, Routledge, London.

Massey, D. (1993), 'Power Geometry and a Progressive Sense of Space', in Bird, J., Curtis, B., Putnam, T., Robertson, G. and Tickner, L. (eds), *Mapping the Futures: Local Cultures, Global Change*, Routledge, London, pp.59–69.

Mohring, H. (1993), 'Maximizing, Measuring, and not Double Counting Transport-Improvement Benefits: A Primer on Closed- and Open-Economy Cost-Benefit Analysis', *Transport Research B*, Vol. 27, Nr. 6, pp.413–424.

Oosterhaven J. and Knaap T. (2002), 'Spatial Economic Impacts of Transport Infrastructure Investments', this volume.

Pieda, (1994), *Heathrow Terminal 5: Assessment of Employment Impact, 1991–2016*, Reading: Pieda, BAA/1201.

SACTRA (1999), 'Transport and the Economy', *Report of the Standing Advisory Committee on Trunk Road Assessment*, The Stationery Office, London.

Storper, M. (1998), 'Regional Worlds of Production: Learning and Innovation in the Technology Districts of France, Italy and the USA', *Regional Studies*, Vol. 27, Nr. 5, pp.433–455.

Swyngedouw, E. (1993), 'Communication, Mobility and the Struggler for Power over Space', in Giannopoulos, G. and Gillespie, A. (eds), *Transport and Communications Innovation in Europe*, Belhaven, London, pp.305–325.

Vickerman, R.W. (2001), 'Transport and Economic Development', Paper presented at the 119th ECMT Round Table, Paris.

Chapter 7

European versus National-Level Evaluation: The Case of the PBKAL High-Speed Rail Project

Rana Roy

The Problem of Evaluation in Cross-border Projects[1]

In the Member States of the European Union, public policy usually mandates that transport infrastructure projects should be selected on the basis of a full socio-economic evaluation, be tested against a hurdle rate of economic profitability and, where necessary, be supported by public funds by virtue of passing this hurdle. That necessity arises where such an *economically* profitable project is nonetheless insufficiently profitable *financially* – a coincidence that is not uncommon.

However, so long as we live in a world of separate nation-states, and even where such states are also Member States of a larger Union, public investment is, by and large, undertaken on behalf of the national taxpayer. It follows that the cost-benefit analysis required to evaluate such investment is also – and correctly so – undertaken from the viewpoint of the national taxpayer.

This generates a particular problem for the European Union in the case of the cross-border projects of the Trans-European Networks (TENs) programme, the importance of which has been stressed time and again by the Union and its member-states but actual progress on which has been painfully slow and uncertain.

The problem is best defined as it was first defined: by reference to the case of the high-speed rail project Paris-Brussels-Cologne-Amsterdam-London, PBKAL.

In any cross-border project where each jurisdiction is principally responsible for the funding of its national section, the evaluation of the project will be fragmented into separate national evaluations of the respective sections. And in order to determine the extent to which their section merits public subsidy *from the national taxpayer*, most national governments will, quite properly, seek to limit their recognition of the section's economic benefits to the share accruing to their own residents. In short, they will calculate their own *national benefit*.

In the UK and France, for example, explicit investment rules served to define the boundary of national benefit. Thus, three streams of benefits were counted:

- the financial return;
- domestic economic benefits: user benefits to passengers making domestic trips on the new international lines – for example, London-Ashford on CTRL, Paris-Lille on LGV Nord – as well as the standard non-user benefits;

- the national share of international economic benefits. This includes, principally, the estimated share of the international consumer surplus which falls to own-residents within its section – an estimate based on a passenger split derived from data sources such as the International Passenger Survey.

In both states, the rules governing the determination of national subsidy explicitly discounted a fourth stream of benefits:

- the 'supra-national' share of international economic benefits: the share of benefits which falls to non-residents. For example: in the case of CTRL, the estimated consumer surplus for French and other international passengers on the British line; in the case of LGV Nord, the estimated consumer surplus for British and other international passengers on the French line.

The result may be schematically represented as follows:

Table 7.1 The Four Streams of Benefits

		Financial return	→	National measure of economic return
Relevant for national subsidy	←	Domestic economic benefits	→	
Relevant for national subsidy	←	International economic benefits to own-residents	→	
Not relevant for national subsidy		International economic benefits to non-residents		Not included

Each national evaluation is perfectly defensible in its approach to the problem of determining the burden to be placed on the national taxpayer. But there is an asymmetry here: the same national taxpayer is unable to evaluate and hence to secure the benefits accruing to his own fellow-residents in the other sections of the same cross-border project. The unintended consequence is that the set of national evaluations will exclude from view no less than half the international consumer surplus.

The point may be illustrated with the aid of a simple two-country example. Suppose that two geographically contiguous countries, A and B, invest in a new cross-border line which generates a time-saving gain of one hour in each of the two national sections: a total of two hours. The benefits are expressed in terms of time-savings only: *ceteris paribus* applies to all the other elements. The required investment cost is one billion Euro in the case of each section: a total of two billion

Euro. The estimated traffic is two million passengers. And in each section, the estimated passenger split between residents and non-residents is exactly 50/50.

In this example, Country A's evaluation of its national section will show *inter alia* a benefit of one million hours of time-saving gains measured against an investment cost of one billion Euro. Country B's evaluation will show just the same. The implicit sum will thus show a benefit of two million hours of time-saving gains measured against an investment cost of two billion Euro. The truth, however, is that the investment in the project as a whole will enable each of the two million passengers to gain two hours of time-savings – thus generating a gain of four million hours of time-saving measured against an investment cost of two billion Euro.

Table 7.2 The Lost and Found Problem

Evaluation of national section *in Country A*	Evaluation of national section *in Country B*	Implicit sum of A + B	Corrected evaluation of *project as a whole*
Investment cost: 1 billion Euro	Investment cost: 1 billion Euro	Investment cost: 2 billion Euro	Investment cost: 2 billion Euro
Time saving: 1 hour	Time saving: 1 hour		Time saving: 2 hours
Resident passengers: 1 million	Resident passengers: 1 million		All network passengers: 2 million
1 million hours of time savings measured against investment cost of 1 billion Euro	1 million hours of time savings measured against investment cost of 1 billion Euro	2 million hours of time savings measured against investment cost of 2 billion Euro	4 million hours of time savings measured against investment cost of 2 billion Euro

This is of course a highly stylised picture. Realistically, the passenger split is likely to vary. Countries with a greater-than-average share of resident passengers on their own sections will capture in their evaluations more than half the international consumer surplus on these sections. But this will be off-set by at least one other country with a less-than-average share. The relevant point is that the sum of 'resident passengers' cannot exceed half the number of total passengers. Hence – and assuming *ceteris paribus* in regard to unit values of time savings and all other elements in the calculation of the consumer surplus – the set of national evaluations will fail to capture half the international surplus.

Indeed, in the case of many of the cross-border TENs, the sum of 'resident passengers' is likely to be *less than half* the total – since passengers from non-participating states are also likely to figure in the network's traffic. Where this obtains, *more than half* the international consumer surplus will be 'lost'.

Thus, *correctly* specified national evaluations of the respective national sections will generate an *incorrect* estimate of the economic profitability of the project as a whole and each of its national sections – and, therewith, an incorrect estimation of the extent of public support which they merit.

At the same time, the variability of returns across the various sections can also serve to distort the overall picture.

On the one hand, a national section with a high economic return may be so placed that a very high proportion of its international economic benefits falls to non-residents so that the national measure of economic return falls below the hurdle rate for attracting subsidy from the national taxpayer. The section is thus jeopardised for want of subsidy even though its full economic return is high.

Per contra, another section with a high economic return may be so placed that it clears the hurdle of financial profitability and proceeds without need of subsidy. The project is thus deprived of a stream of subsidy corresponding to that stream of economic benefits even though the project as a whole may be in need of it.

The general problem outlined above was strikingly manifest in the case of the PBKAL. Prior to the identification of the problem in *Lost and Found*, the sum of deliberations on national subsidy did indeed result in the non-recognition of more than half the international consumer surplus. In most states, the national measure of economic return seemed insufficiently high to merit the degree of public subsidy required for the national section. In the one state where the national measure alone did suffice to justify the investment – France – the *ex ante* evaluation also suggested that the financial profitability of the section was high enough to enable it to proceed without need of public subsidy. Consequently, subsidy was not provided.

Not all international transport infrastructure projects necessarily encounter this problem. Road projects are normally evaluated on the basis of the benefits and costs to all the traffic in the relevant network irrespective of the nationality of the driver, vehicle or goods. However, similar problems to those analysed here may arise where border-crossing roads are appraised at national level if the value of the whole route is greater than the sum of the parts.

Nor is the case described above the only feasible one. It might easily be that the geography is such that for a project which is overall worthwhile in cost-benefit terms the 'correct' benefit to one participating country is less than the cost to that country. In such a case a side-payment to the losing country would be required. So, the case under consideration is not a universal case in multi-country project appraisal, but it is an important real case.

A Step to a Solution: The Second Level of Evaluation

At one level, the problem and its solution can be expressed very simply. The problem is that national evaluations will tend to exclude a large stream of benefits as irrelevant: the supra-national benefits. The solution therefore is to restore these benefits to view, in one way or another.[2]

At a deeper level, it needs to be understood that the solution to the problem depends on the presence of a decision-maker with an interest in, and responsibility for, recognising and securing this stream of supra-national benefits.

If the exclusion of these benefits were simply the result of an error in the national evaluations, the solution would consist simply in undoing that error. Thus, each state could amend its evaluation so as to capture the full value of the international consumer surplus generated on its section. Indeed, the recommendation that each of the PBKAL Member States *should* amend their evaluations accordingly was a not uncommon initial response to our identification of the problem.[3]

In our analysis, however, the national evaluations are not *erroneous* but rather – and necessarily – *incomplete*. And the response described above is ill-advised for several reasons.

First of all, the exclusion of non-resident benefits in determining the subsidy from the national taxpayer is a perfectly defensible procedure. If all states were to include all international benefits in determining the national subsidy, those states with a higher-than-average share of non-resident traffic would end up making a gift to the others. And whilst it is their prerogative to make such a gift, it is not their obligation to do so. Nor are the other states entitled to demand this gift as of right.

Of course, a gift can sometimes make sense. In some cross-border projects – consider, for example, the Øresund Link, a project involving only two states with long-established patterns of co-operation and generating what is apparently a reasonably even spread of benefits – mutual recognition of each others' benefits might well appear to be the most efficacious solution.

Nonetheless, there is a well-founded presumption that subsidies should be transparent. Just the same applies to international cross-subsidies. If participating states were to choose to subsidise each other's benefits, that subsidy ought to be calculated by agreement rather than being the involuntary outcome of the failure to distinguish between the various streams of benefits.

The general form of the solution to the problem of evaluation in cross-border projects is thus to recognise the distinction between the streams of benefits identified above and *then* to re-integrate the excluded fourth stream. It is not a matter of *amending* erroneous national evaluations. It is rather a matter of *complementing* incomplete national evaluations with a second level of evaluation: a supra-national evaluation.

The exact form of the evaluation to be applied will depend on the nature of the project as well as the identity of the decision-maker. A single project between two wholly independent states to be co-funded by mutual agreement will not necessarily require the same treatment as a programme of multiple projects between several Member States of a larger community. Moreover, the weight of the presence of third parties – passengers from non-participating states – will also bear on the question of the form of the evaluation.

In the case of the cross-border projects of the TENs programme, it is clear that the European Union has an interest in recognising and securing these supra-national benefits in the national sections of any given project – and in establishing for itself a true measure of the economic return on the project as a whole. Where there are excluded benefits these are overwhelmingly benefits to the residents of the Union, be they from participating or non-participating member-states. The exclusion of these benefits makes it impossible to judge the value to the Union of any given project – or the relative value of one TENs project over another. Furthermore, their exclusion ensures that the profitability of cross-border projects will be systematically under-estimated relative to purely national projects. In sum, the Union has an abiding interest in ensuring that the problem *is* solved.

In the case of the PBKAL, therefore, the solution was to produce for the Union a European evaluation to complement the national evaluations of the participating Member States – one which was intended to be the first in a series of European evaluations for the full spectrum of Trans-European projects.

The evaluation itself presupposes no more than fully specified national evaluations. In each national section, the data for calculating the fourth stream of benefits are the same as those for calculating the third stream. Thereby, we can first establish the full economic return on each section – but distinguishing between its national and supra-national components. Secondly, by averaging the national inputs on a weighted basis, we establish the full economic return for the project as a whole. Derivatively, the procedure thus establishes the 'community component' for the project as a whole:

Table 7.3 The Lost and Found Solution

	Cumulative rate of return (rounded)
Financial return	
Add domestic economic benefits	
Sum of above	
Add international economic benefits to own-residents	
National measure of economic return	X
Add international economic benefits to non-residents	
Corrected measure of economic return	X*
Community component of economic return	(X* – X)/X*

Establishing this community component provides a rational basis for determining the appropriate level of Union subsidy where necessary.

To be sure, this does not mean that Union subsidy will always be necessary. Union subsidy is a sub-set of public subsidy. If it is to be guided by the same principles of public economics that inform the determination of public subsidy at the national level *plus* the additional principle of subsidiarity, then the following set of criteria will obtain:

- the need for a Union subsidy in cross-border TENs projects will arise only in those cases where the combination of financial return and national public subsidy is insufficient to enable the project to be completed;

- such projects will merit subsidy only in those cases where (1) the project's corrected measure of economic return is sufficiently high and (2) the community component of that economic return is also sufficiently high;
- both targets need to be met. If (1) is not met, it indicates that the project does not merit public subsidy – even though a high proportion of its relatively low benefit to society takes the form of supra-national benefits. If (2) is not met, any Union subsidy would fail its purpose of securing significant supra-national benefits and serve only to secure the national benefit of particular member-states.

As is shown below, the PBKAL met both these targets.

The Case of PBKAL: Main Results and Conclusions

It is unnecessary here to report the full set of results of the new European evaluation of the PBKAL established in *Lost and Found* or to rehearse all the steps by which that evaluation was established (Roy, 1995, and EC, 1995). Nor is it necessary for the purpose of the present discussion to revise or update those results in the light of subsequent developments. For the main results and conclusions of the 1995 work suffice to show their general and continuing relevance for the TENs programme.

By 1995 the French section of the project had been completed. But the UK, Belgian and Dutch sections were in severe financial difficulties. Given the limits to the level of national subsidy suggested by their respective national evaluations, none of these Member States were in a position to complete their national sections. The relatively small German section had not yet been fully evaluated.

Following the necessary adjustments to ensure that each of the three main national evaluations in question were fully consistent with each other[4] – and having reduced the German contribution to nil so as to ensure that we erred on the side of underestimating rather than overestimating the project's overall profitability[5] – the new evaluation showed the following results for each national section and for the project as a whole:

Table 7.4 PBKAL: Modelling the Community Component

	UK	Belgium	NL	D	Total
Investment cost (in million Euro)	3,800	3,700	3,400	700	11,600
Proportion of costs	32.8%	31.9%	29.3%	6.0%	= 100%
Financial return	4.2%	2.6%	1.5%	0.0%	
Weighted share	1.38	0.83	0.44	0.00	
Weighted average					2.7%
Add domestic economic benefits					
Sum of above	6.0%	4.2%	3.2%	0.0%	
Weighted share	1.97	1.34	0.94	0.00	
Weighted average					4.3%
Add international economic benefits to own-residents					
National economic return	8.4%	5.6%	6.7%	0.0%	
Weighted share	2.75	1.79	1.96	0.00	
Weighted average					6.5%
Add international economic benefits to non-residents					
Corrected economic return	10.4%	8.6%	9.2%	0.0%	
Weighted share	3.41	2.74	2.70	0.00	
Weighted average					8.9%
Community component of economic return	19.2%	34.9%	27.2%	N/A	27.0%

For ease of reference, the overall results for the project as a whole may be presented as follows:

Table 7.5 PBKAL: Results for the Project as a Whole

	Cumulative rate of return (rounded)
Financial return	2.7%
Add domestic economic benefits	
Sum of above	4.3%
Add international economic benefits to own-residents	
National measure of economic return	6.5%
Add international economic benefits to non-residents	
Corrected measure of economic return	8.9%
Community component of economic return	(8.9 – 6.5)/8.9 = 27%

Taking the project as a whole – and applying as we did an economic rate of return of eight per cent as the appropriate hurdle rate[6] – Table 7.5 shows that:

- reflecting the project's highly international character, the community component was of a very high order: 27 per cent of the full economic return;
- with it, the economic profitability of the project, at 9 per cent, passed the hurdle rate; without it, at 6.5 per cent, it did not.

Turning to the national sections as reported above in Table 7.4, it can also be seen that:

- each of the national sections in question displays a very high community component;
- with it, each of the national sections passed the hurdle rate; without it, most did not. In most cases, the national measure of economic return was simply insufficient to justify the commitment of public funds to the completion of these lines.

In turn, this provided a rational explanation for the impasse at which the project had arrived – in particular, the crisis in the Belgian section which had initially prompted the convening of the PBKAL Working Group of the participating Member States, the European Investment Bank and the European Commission. The point was not that, geographically, Belgium held what property developers call the 'ransom strip'. The point was, rather, that in terms of public economics, a community component at 35 per cent of the full return on this section was too large a gift for the rest of Europe to demand of the Belgian taxpayer. In terms of Net Present Values (NPVs), the fourth stream of benefits on the Belgian section amounted to 2.4 billion Euro – a very large sum indeed in comparison to the Belgian foreign aid budget![7]

In sum, the new European evaluation showed that the PBKAL was a good project in terms of its full economic return, good enough to justify the requisite public subsidy – but good enough *only by virtue of* the community component of its economic return.

In view of these findings, it was possible to conclude, as we did, that the project justified Union intervention to secure its large supra-national benefits – and hence Union subsidy up to the then-prevailing Ecofin-stipulated limit of 10 per cent of total investment cost, that is, around one billion Euro.

The Follow-up: The Project, the Top-up and the TENs

In terms of its *immediate* follow-up, the new evaluation led to a successful outcome by any measure. Thus:

- from an antecedent situation of conflict between the participating Member States, all the Member States and the Union institutions represented in the Working Group were able to arrive at a consensus – as embodied in its Final Report (EC, 1995) – on the nature of the problem and of the solution required;
- the analysis was adopted by the European Commission and applied to its deliberations on the TENs in various subsequent reports (EC, 1997, and EC, 1998);
- most importantly, the new consensus led to a series of subsidy decisions – Union subsidy as well as appropriate adjustments to national subsidies – that enabled the project to proceed.[8]

In short, the PBKAL was saved – immediately in the case of the vital Belgian section. And notwithstanding subsequent delays and disappointments elsewhere, it can now be said that the project as a whole is likely to be completed.

Moreover, *Lost and Found* enabled the Commission to build a successful case for an expansion of the TENs budget so as to secure the supra-national benefits of *all* the cross-border projects – beginning with the injection of one billion Euro as the first instalment of what is projected to be a multi-billion Euro 'top-up'.

Nonetheless, there is a larger and long-term sense in which the outcome of the 1995 initiative has not as yet been the success that it was expected to be.

The European evaluation of the PBKAL was conducted – and the support of the non-participating Member States for the Union-led rescue of the PBKAL was secured – on the understanding that this form of evaluation was to be extended to the TENs programme as a whole. This was held to be especially critical in the case of the Brenner project, in which Austria found itself in much the same economic position as Belgium in the case of the PBKAL. It was also seen as a vital element in the expansion of the TENs beyond the existing Member States: to welcome the applicant countries with demands for *de facto* 'gifts' of supra-national benefits would appear to be even more inappropriate than the demand that had been placed on Belgium.

None of this has come to pass. *Lost and Found* has been endorsed often enough in Commission reports; but it has not been applied. The top-up has been secured at least in part; but it has not been put to its intended use. The allocation of subsidies to TENs projects remains the uncertain outcome of the play of political bargaining. And progress on the TENs continues to disappoint – thanks in part to the failure to overcome the problem of evaluation identified in *Lost and Found*.

The Lacuna in the Original and the Corrective Thereto

Against this background, it must now be asked whether there is not a lacuna of some kind in the original analysis and, if so, what if anything can be done to correct it.[9]

There *is* a lacuna in the original analysis (both in Roy, 1995, and in its adoption in EC, 1995). What is presented here is a new form of evaluation applied to a single project – along with an *appeal* to the Union to intervene so as to secure its identified benefits and the comparable benefits of comparable other projects. What is missing is the prescription of a *determinate process of non-discretionary decision-making* by which the new supra-national European evaluations are mandated and the newly identified supra-national European benefits are secured. So long as such a process is not agreed upon and instituted, some projects will be rescued by political discretion but others will not – precisely as a matter of discretion.

Contrast the case of purely national projects. In most EU Member States, a more or less comprehensive evaluation of all projects of any serious size is mandatory. Final decision-making does contain elements of discretion but is at least informed by mandatory evaluation as well as being partially constrained by formal decision rules and hurdle rates. In France in particular, there has long prevailed a more or less binding decision-rule turning hurdle rates into trigger rates: implement all infrastructure projects with an economic return of >8 per cent; reject all those of <8 per cent. And quite irrespective of the element of discretion, the process of decision-making is everywhere an established and determinate process – not an ad hoc or happenstance occurrence.

The case for instituting such a process in regard to the TENs – indeed, doing so more comprehensively than currently obtains for national projects in the Member States – is likely to be strengthened by the findings of recent scientific research as well as by the outcomes of recent political initiatives.

Thus, and specifically with regard to cross-border projects, recent research has shown (Bassanini and Pouyet, 2000) that the absence of agreed and optimal levels of public subsidy for a given project taken as a whole can make it rational for decision-makers on its national sections to set both prices and subsidies at levels which impose sub-optimal outcomes on others. This extends the analysis from the singular question at issue in *Lost and Found* – the question of whether a given cross-border project or one or more of its national sections would fail to be implemented in the event of its supra-national benefits continuing to be excluded from view – to the larger territory of the possibilities and consequences of free-riding in the case of existing projects. Hence, it discloses the larger price that will be paid by society for the failure to solve this particular problem.[10]

More generally in regard to decision-making on infrastructure projects, recent work for the ECMT has made the point (Roy, 2000, and ECMT, 2000) that the need for a binding decision-rule on investment will become yet more urgent in the course of, and as a means to, implementing efficient pricing. For a pricing regime based on short-run marginal cost needs to be complemented – in the long run – precisely by such an investment regime. Without it, the revenues from congestion pricing will increasingly be perceived not as a derivative of the act of correcting prices but rather as the targeted outcome of a deliberate under-provision of infrastructure. Without it, therefore, the consensus supporting the pricing regime and hence the pricing regime itself would face the threat of collapse.

In response to the recent ECMT report, the Prague meeting of the ECMT member countries mandated further work on decision-making as a priority item. This may well provide a forum within which the case outlined above – the case for a determinate process of non-discretionary decision-making on the TENs, both within and beyond the EU – could be advanced.

Once such a process is agreed upon and instituted, the details – the precise mechanisms to be deployed, the extent of their resourcing and the nature of their empowerment – will still need to be settled. An important question to be raised in this regard is whether or not *new* institutional mechanisms are required – funded perhaps by a re-allocation of the multi-billion Euro TENs 'top-up' that has yet to be put to its intended use.

The first priority however is for the relevant decision-makers to agree upon and institute the process itself. Without that agreement, breakthroughs in evaluation will remain mainly an item of academic interest, the rescue of particular projects will remain a matter of discretion, and the larger part of the benefits identified in *Lost and Found* will surely be lost again.

Notes

1 The core of the analysis and results presented in this paper is derived from my work, *Lost and Found: The Community Component of the Economic Return on the Investment in PBKAL*, originally published by the European Commission/PBKAL Working Group as an annex to *PBKAL: Final Report on the High-Speed Rail Project Paris/London-Brussels-Cologne(-Frankfurt)/Amsterdam*, Brussels, 1995 – and subsequently published separately, unabridged and with a Foreword by Henning Christophersen, as an ECIS Report, Rotterdam, 1995. References to *Lost and Found* and to *PBKAL: Final Report*, are given in the text as Roy, 1995, and EC, 1995, respectively. It only remains to add that my acknowledgements in the original – to the Working Group as a whole, to its many individual participants, and especially to my colleague and co-author of the detailed model we developed for the exercise, Dr. Lawrence Harrell – should be read as carried over to this paper.

2 This point holds true, notwithstanding anything else there is to be said on the matter. Thus, the problem cannot simply be dismissed by noting that, in practice, no evaluation can hope to capture all possible impacts. The benefits identified here are core socio-economic benefits – a part of the consumer surplus that the European 'common

framework' on methodology defines as one of the 'mandatory' rather than 'discretionary' impacts to be evaluated (ITS, 1995). They are also very large – larger by far than those additional benefits that have been identified in recent discussions as escaping the net of conventional cost-benefit analysis (ECMT, 2000 and Roy, 2000).

3 Roll, 1996, for some instances of this response.

4 *Inter alia*, consistency required that the base case in the national evaluations of the UK, Belgian and Dutch sections incorporated the French section from the outset; that each of these national evaluations included in their respective base case the successive entry of the other national sections, according to the common calendar adopted by the Working Group; and that the period of calculation in the evaluation of each of these sections terminated at a common point: the year 2035. Subsequent developments have ensured that the UK section – the CTRL – will not be completed in accordance with the common calendar.

5 At the time, the sole input supplied by DB AG and the German Transport Ministry on their national section was an estimated financial return of 0.03 percent. In our model, all rates of return on the German section were set at zero.

6 Roy, 1995, for the background discussion on the choice of this hurdle rate.

7 NPVs for each stream of benefits and for each national section were reported in the original study (Roy, 1995, Attachments 5, 6, and 7). NPVs for supra-national benefits amounted to around £1 billion and NLG 2.5 billion in the UK and Dutch sections, respectively. However, for a number of reasons, including the fact of widely differing discount rates among the participating Member States as well as simply ease of reference, the results for the project as a whole were given in rates of return only.

8 The actual levels of subsidy are another matter. However, I had always taken the view that Union intervention, even if insufficient, would act as a sufficient catalyst for new subsidy decisions on the part of the participating Member States. As *Lost and Found* put it, 'it would suffice to prompt the necessary upward adjustment in the sum of national subsidy. By discharging its own responsibilities, the Union would be well placed to oblige all parties to discharge theirs' (Roy, 1995).

9 As early as 1996, in a report for the UIC on *Benefits of Rail Infrastructure Projects* (Roy, 1996), I had identified two problematic issues in the analysis in *Lost and Found:* First, the current TENs budget line does not permit the requisite level of Community funding – the PBKAL alone merits one half of the resources available – and the Union institutions seem temporarily unable to resolve the impasse on the TENs budget. A second, more intractable barrier is that the Commission services at present possess neither the technical capability nor the institutional authority to conduct the requisite evaluations and design the requisite subsidy arrangements to a standard acceptable to their governmental, commercial and financial partners in the Member States.

The first issue has been resolved – even if the final execution of the solution still lies ahead of us. The second however remains a problem and an intractable one at that. What follows situates it as one element of a larger problem: the absence of a determinate process of decision-making. The process must first be agreed upon before it can be appropriately supported and empowered.

10 The insightful paper by Bassanini and Pouyet is in fact part of a larger research programme supported by RFF and the French Ministry of Transport which should lead to further insights on the nature and extent of the price of failure on this issue.

References

Bassanini, A. and Pouyet, J. (2000), 'Optimal Access Charges with Interconnected Railroad Network', Paper presented to Railroad Conference on New Developments in Railroad Economics, Infrastructure Investment and Access Policies, Paris (La Defense).

EC (1995), European Commission/PBKAL Working Group, PBKAL, *Final Report on the High-Speed Rail Project Paris/London-Brussels-Cologne(-Frankfurt)/Amsterdam*, Brussels.

EC (1997), European Commission/High Level Group on PPPs, *Report of the High Level Group on Public-Private Partnership Financing on Trans-European Network Transport Projects*, Brussels.

EC (1998), European Commission, *Fair Payment for Infrastructure Use, A Phased Approach to a Common Transport Infrastructure Charging Framework for the EU*, White Paper, Brussels.

ECMT (2000), *Assessing the Benefits of Transport*, ECMT Report [pre-publication version], Paris.

ITS (1995), Institute for Transport Studies (University of Leeds) *et alii*, *Cost-Benefit and Multi-Criteria Analysis for New Infrastructure in the Field of Railways*, prepared for the European Commission, Brussels.

Roll, M. (1996), *The Trans-European High-Speed Rail Networks – A Socio-Economic Analysis. Case Study, The High-Speed Rail Project Paris-Brussels-Cologne-Amsterdam-London (PBKAL)*, Vrije Universiteit Brussel, Masters Thesis, Brussels.

Roy, R. (1995), *Lost and Found, The Community Component of the Economic Return on the Investment in PBKAL*, originally published by the EC/PBKAL Working Group as Annex to *PBKAL, Final Report on the High-Speed Rail Project Paris/London-Brussels-Cologne(-Frankfurt)/Amsterdam*, Brussels, – and subsequently published as an ECIS Report, Rotterdam.

Roy, R. (1996), *The Benefits of Rail Infrastructure Projects*, ECIS Report for the UIC, Rotterdam.

Roy, R. (2000), 'Means and Ends, Cost-Benefit Assessment and Welfare-Maximising Investment', *Assessing the Benefits of Transport*, ECMT Report [pre-publication version], Paris.

Chapter 8

Welfare Basis of Evaluation

Marco Ponti

'Evaluation' among other things means the elaborations performed by technicians in order to help politicians in their decision-making activity. No assumption of 'technical dominance' is therefore implicit in the process: decision-making remains a *political* procedure. Furthermore 'evaluation' it is *not* assumed as limited to infrastructure projects but it is related with any transport policy, including regional, national and European planning.

'Welfare' implies its standard economic meaning: it is composed by: a) net social surplus (willingness to pay or willingness to accept, minus the social costs of the resources consumed); b) distribution effects, i.e. gains and losses of different social groups.

It is suggested to combine some innovation in welfare economics with an updated *public-choice* approach. Examples of 'capture' of decision-makers by interest groups are outlined with special reference to European policies to create a single market for transport. Cost benefit analysis (CBA) is recommended to limit as far as possible 'discretionary' values directly given by the decision-makers.

In the areas where discretionary values are considered inevitable – for example, re-distribution of income, environmental standards and landscape/aesthetic values – it is argued that 'switch' economic values can and should *always* be made explicit. This means that the opportunity cost of achieving one objective (e.g. to reduce environmental damage) against the other (to increase the lot of a low-income group) can and has to be calculated in order to guarantee *consistency* within the decision-making process and to reduce discretionary attitudes of politicians. A simplified example is then presented to estimate the opportunity costs of reaching Kyoto targets for reduction of CO_2 emissions in road transport by means of comparing the social cost (variations in surplus) of reaching such targets through alternative policies.

Improvements to the standard CBA practice are separately discussed in three areas:

- *marginal opportunity cost of public funds*: to take into account that, if a project/a policy increases the state deficit, this has a *measurable* and *negative* impact on welfare;
- *recent literature developments on social discount rate*: option value (to take into account the degree up to which transport investment can be considered reversible) and environment-related risk-aversion;

- *market failures in transport-using sectors*: to point out possible distortions (in assessment) due to unrealistic assumptions concerning land markets and local monopolies.

Implications of a 'Public Choice' Approach on the Evaluation Tools

Assumptions

Standard cost benefit analysis (CBA) assumes that the key objective of government is to maximise net social welfare (consumer's and producer's surplus, corrected by some progressive redistribution goal), and provides measurements and guidelines for this maximisation. Multi-criteria analysis (MCA) considers governments 'benevolent princes': it then asks decision-makers to represent the public interest by giving a weight to (the desirability of achieving) different objectives. With a solid base of evidence, public-choice theory (PCT) has elaborated on the assumption that politicians and bureaucrats, as the *homo economicus* of the classical approach, have their own self-interested, egoistic objectives, basically related to 'winning power, exercising power and hanging on to power'.[1]

Recognising that elected governments often have a short-term, egoistic 'hidden agenda' (e.g. re-election), and that also their long term objectives may be distorted (e.g. conceived to favour some social groups at the expense of another, but for partisan reasons), Cost-Benefit Analysis seems the most effective tool, within an updated 'public-choice' approach, for *limiting* as far as possible 'discretionary' values given by decision-makers.

As a matter of fact, the only areas where discretionary, 'political' values should be considered inevitable are a) income redistribution (a highly critical issue at every level of decision-making), b) environmental standards (increasingly set at international level) and c) landscape or architecture-related aesthetic values (quite a crucial issue in local planning).

Redistribution objectives are critical in two ways. Actually, the *direct* distribution of costs and benefits to different social groups of specific projects/ policies/ public spending programmes is not too difficult to evaluate. If these impacts are quantified, the transparency of the decision-making process is increased, and consistency can be better guaranteed, even if the actual distributive 'weights' remain strictly, and correctly, within the political sphere. But the *total* distributive impact of any public action is strictly related also to the real distributive impact of the taxation regime. The more progressive the taxation regime, the smaller the level of fiscal fraud, the more redistributive the impact even of an income-neutral project. But for example, if labour is heavily taxed, as in many European countries, transport projects that increase the public deficit, and therefore the overall fiscal pressure, may raise some doubts about their final distribution impact (see also the opportunity cost of public funds). But even excluding the taxation issue, the need for *explicit* and *quantified* distributive analysis of costs and benefits seems consistent with a PCT approach to evaluation.[2]

The 'Capture' of Decision-makers by Incumbents

In the last decade, Member States have been transferring substantial regulatory powers to level the field for competition in the single market to the Commission. A steady pace in formal adoption of European Directives has created widespread hopes of a reshaping of national transport industries. But attempts to create a single market for transport services have proved more complex in industries where the state presence was dominant, such as railways, infrastructure concessions, air services and local public transport.

In continental Europe, difficulties and delays in these areas are indicating that incumbent monopolistic industries are still in a position to resist liberalisation and to influence national transport policies, including by means of 'negotiating' the amount and destination of public funds. These difficulties would also indicate that 'capture' (known in more plain language as 'cosy relationships') of national governments by incumbent monopolists has had an influence on reducing the effectiveness of European Directives.

Also the outcome of attempts to liberalise rail and air services in the Union would confirm well-known arguments from the literature, as well as the available evidence, that *partial* liberalisation seldom works, due to explicit or hidden cross-subsidisation of liberalised activities by monopolistic ones:

- *railway liberalisation* and open access were considered key options for revamping a declining sector that generates egregious financial costs to Member States, and for promoting integration of rail-based services in the single market. As the implementation of the liberalisation policy has been delegated to the national railways companies, i.e. the incumbents, small wonder that, ten years after adoption of Directive 440/91, a new generation of rail companies was not able to enter the market for rail services, and, still worse, cartels developed;[3]
- a similar fate happened to continental European *civil aviation*: 'flag' carriers are still dominant, with high tariffs and blatant evidence that this dominance has little to do with their capacity for competitive excellence. In such a context, national governments are protecting flag companies also by opposing any attempt to liberalise airport slots and intercontinental services. These are the two main reasons also behind the weak powers of pro-competition authorities in protecting consumers against the potential cartel aspects of international strategic alliances, which are developing at a steady pace among air carriers as a result of partial liberalisation;[4]
- as the Commission stepped into *local transport* (see Green Paper *The Citizen Network: fulfilling the potential of public passenger transport in Europe*, 1995), subsequent attempts to make public transport more efficient by means of introducing competition, even only for the allocation of public service contracts, faced severe resistance. For example, the number and type of possible exemptions introduced to the last proposal for European Regulation[5] is so wide as to make it easy to imagine how incumbent operators are able to protect the 'status quo'.

Further examples in successful 'capture' of policy-makers by special interest groups can include:

- reduced taxation by national governments to protect national road haulage industries from international competition and/or variations in relative prices; more recently, as crude oil was approaching $35 per barrel, road hauliers were also forcing many Member States to forget environmental targets and to introduce compensations for high fuel taxation;
- subsides still widely granted to incumbent national rail operators in the process of liberalisation of rail freight transport in the single market;
- the interests of the European motor industry in continuing to produce road vehicles able to reach over 200 km per hour may explain reluctance in the enforcement of existing speed limits or in measures to reduce them (as a mean to increase road safety and to reduce air pollution);
- insufficient determination of local governments toward charging for the use of scarce road capacity (road/parking pricing) and or/to limit the use of private vehicles in congested urban areas often reflect resistance from 'vocal' groups of residents and retailers.

Switch Values

Estimating 'switch' values, that is the value of a given variable that may turn a profitable project into a less profitable one, is commonly used to carry out sensitivity analysis in CBA assessments. Other than for estimating the Internal Rate of Return (the rate of discount that equalises discounted flows of costs and benefits) switch values can be estimated for any critical parameter in a given project, from the level of demand to shadow prices.

In dealing with non-quantified values, the estimation of a switch value can be considered as a means to persuade the decision-maker to put a value on something (or at least to say that the value is greater, or less than a given value) in a specific context. As pointed out in the critical revision of the 1968 OECD Manual of Industrial Project Analysis carried out by Little and Mirrlees (1974), when more than one variable is considered (income distribution or environmental harm in the quoted example), and more than one decision-maker is involved, estimating switch values does not reduce the need for quantification of economic values. This quantification is still needed in order to reduce shortfalls implicit in MCA assessment, including in terms of possible inconsistency in the judgement of an individual decision-maker and/or among different decision-makers.

Transport and the Environment: An Example in the Estimate of 'Switch Value' (i.e. Opportunity Cost)

The debate on transport externalities is a good example of arguments produced by conflicting political positions. In the presence of many unsolved problems on how to estimate the value of external costs caused by transport activities, the discussion that followed EU consultation papers on fair and efficient pricing (CEC, 1995) and infrastructure charges (CEC, 1998) can be broadly associated with two symmetric positions each one developed by political groups and vested interests.

From the point of view of those representing road transport, key arguments are presented by means of (either static or dynamic) comparisons between (in general, 'rock bottom') estimates of external costs generated by road transport with the (huge) amount of fiscal revenue it generates. This position is pointing out relevant problems in terms of fairness. Being based on the notion of average costs and average fiscal revenues, it focuses on fairness, and does not consider the use of efficient (marginal) price signals as a instrument to promote efficiency.

On the other extreme, the arguments produced by the supporters of public transport (as a rule), underestimate the part of costs already internalised by road transport (either through fiscal means or directly), and ignore distribution implications. For example, if the 'polluter pays' is accepted as an equitable principle, it is also equitable that activities (and social groups) that are not generating damages to third parties are reimbursed, if an efficient price is charged (as in the case of road congestion). Furthermore, it tends to be forgotten that: a) internalisation measures do have a social cost (both in terms of the opportunity cost of the fiscal pressure, in case of taxation, or in terms of the surplus losses associated to 'command and control' policies); b) subsidies to public transport are themselves generating inefficiencies, in terms of distorted price signals to users (for example, accelerating urban sprawl phenomena), i.e. subsidies have some characteristics of an externality.

An example is given below for Italy on how to estimate the opportunity cost of alternative policies to achieve the same environmental target by means of measuring variations in social surplus. The estimate is highly simplified and not dynamic in nature.[6]

Given uncertainty in appointing monetary values to external costs, a threshold is set, above which external costs are assumed to be infinite: this constraint is equal to the Kyoto target for reduction of CO_2 transport emissions by 20 per cent by the year 2010. The analysis is limited to CO_2 generated by private cars. Using broad figures for the Italian context (including in terms of present modal share) estimates are made for a fleet of 30 million cars consuming some 30 billion litres of fuel per year (an acceptable 'proxy' for CO_2 emissions).

Within a 'stay-on-the-safe-side' approach, prudential assumptions are used to estimate and compare variations in social surplus as a result of adoption of two alternative strategies;

- to shift passengers from cars toward collective transport modes;
- to modify the car fleet by means of introducing less polluting technologies.

Without discussing specific measures in each package, nor cause-effect patterns, both strategies are assumed to be effective in order to reduce CO_2 emissions by 20 per cent. It is nevertheless assumed that both strategies are implemented by means of introducing price signals (direct charges and taxation) and *not* by means of imposing physical and/or technical constraints. Surplus losses estimated in Figure 8.1 are therefore estimated on a marginal basis (people who leave the car for public transport are those who have a lower utility from the use of car).

Since the aggregated demand function is not known, the surplus lost by those that shift to public transport is estimated on the basis of its inferior limit: fiscal imposition both on fuel and on other variable costs of operating vehicles (such as maintenance, lubricants and road tolls), as indicative of the minimum willingness to pay of the car users above the costs of the resources consumed. Using values that are indicative of the present situation in Italy, this is approximated to be equal to one Euro per litre of gasoline. The surplus loss implicit in the modal shift is therefore 6 billion Euro (30bn x 1 x 0.20).

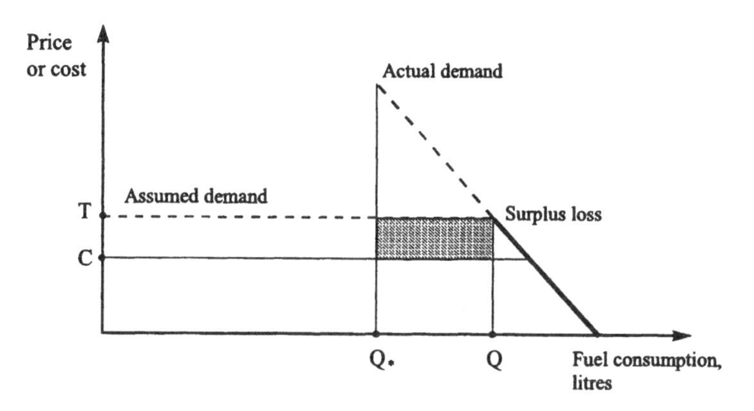

Figure 8.1 Modal share and fuel consumption

Q = present consumption of fuel
Q• = consumption reduced by 20 per cent
C = economic costs of travelling by car
C – T = total fiscal imposition (on fuel etc.)

Assuming (for prudential reasons) that public transport systems do have spare capacity, the marginal costs of the modal shift are estimated at 0.05 Euro per passenger-km. Suppressed costs of using cars are not taken into account (this amount is large, but its value is implicit within the surplus loss evaluation).

As 20 per cent of 540 billion passenger km have to be served by public transport (30 million cars x 12,000 km a year x 1.5 passenger per car) at a cost of 0.05 Euro each, some 5.4 billion Euro a year are to be spent to save 6 billion litres of fuel (30 billions x 0.20). Adding to this amount the surplus loss, estimated above at about 6 billion Euro, we reach a unit opportunity cost in the order of two Euro per litre of fuel saved.

We then assume that the same amount of fuel can be saved by means of introducing hybrid cars.[7] If additional costs of vehicle substitution are assumed at 2,500 Euro per vehicle, we reach a total capital cost of 7.5 billion Euro per year (2,500 Euro divided by 10 years of vehicle life x 30 millions vehicles), the unit cost of saving 6 billion litres of fuel will be in the order of 1.25 Euro per litre.

When comparing this value with the two Euro per litre in the modal shift scenario, we have to remember that the estimates are made assuming a demand curve which is infinitely elastic. Any inclination of the shape of that (flat) curve in Figure 8.1 would increase significantly the surplus losses estimated for the modal shift case.

Furthermore, while both strategies generate social benefits in terms of lower fiscal pressure (unless the lost fiscal revenue is raised from other taxation sources), the estimate does not take into account that the modal shift scenario may require additional subsidies (see also below).[8]

Advancements in CBA[9]

Evaluation of the Marginal Opportunity Cost of Public Funds (MOCPF)

Historically, financial and economic analysis have been kept *separate*, sometimes assuming financial viability as a pre-condition for economic viability (and in fact this pre-condition is generating the Ramsey-pricing scenario, in which budgetary constraints 'drive' the pricing criteria).

This picture has to be changed in Europe by the observation that one financial objective has been assumed as specially relevant for welfare: fiscal pressure reduction. Fiscal pressure, linked with public deficit (and debt), is in fact assumed to *reduce* welfare, if it reaches extreme levels. It is a kind of 'limit' to Keynesian policies (probably generated by asymmetric public spending and recovering in expansion and recession periods, mainly for short-run consensus motivations). This observation in turn defines the 'Maastricht constraints', that are obviously highly differentiated among countries with different public debts.

Evaluation has then to take into account the shadow cost of financial constraints (every constraint does have a shadow cost), and in this way a precise 'bridge' has to be built between financial and economic analysis. Both the Ramsey approach and traditional cost-benefit scenarios are therefore no longer valid.

Let us examine the pricing problem for a transport infrastructure. Short-run social marginal cost pricing (as, for example, recommended by the High Level Group) cannot be accepted any longer in traditional terms. The efficient price, p has to take explicitly into account both deadweight welfare losses (DWWL, area A in Figure 8.1) of the total cost recovering price (P_2) *and* the marginal opportunity cost of public spending necessary to finance total investment cost (MOCPF; the welfare losses as a percentage k of the subsidies needed – areas B in the Figure).

In order to minimise the total welfare loss, the efficient price has to satisfy the equation:

$$\frac{d(MOCPF)}{dp} + \frac{d(DWWL)}{dp} = 0$$

With MOCPF a *direct* function of price, and DWWL an *inverse* function, the efficient price:

$p^* = P_1(1+k)$

is therefore estimated assuming a constant-elasticity demand function ($q = 1/p$) as detailed in Box 8.1.

Figure 8.2 Efficient Infrastructure Pricing with Explicit Opportunity Cost of Public Funds

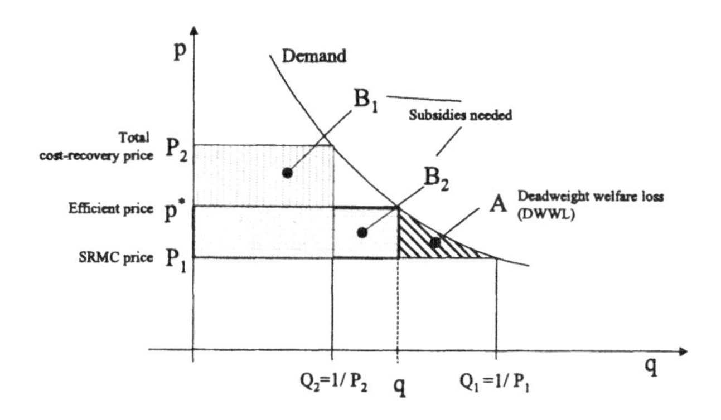

Box 8.1 Estimation of the efficient price[10]

As the function $\frac{d(MOCPF)}{dp} + \frac{d(DWWL)}{dp} = 0$ can be re-written into

$$\frac{dA}{dp} + k\frac{dB}{dp} = 0$$

$$A = \int_{P_1}^{p} (1/p)dp + P_1/p - 1 = \log p - \log P_1 + P_1/p - 1$$

$$B_1 = 1 - P_1/P_2$$

$$B_2 = 1 - P_1/p$$

$$B = B_1 - B_2 = P_1(1/p - 1/P_2)$$

$$\frac{dA}{dp} + k\frac{dB}{dp} = 1/p - P_1/p^2 - kP_1/p^2 = 0$$

$$p^* = P_1(1+k)$$

The MOCPF is also relevant in the analysis of environmental issues. If environmental taxation (fuels taxes, road pricing, etc.) is forced by this opportunity cost to *substitute* other forms of taxation, for example labour taxes, positive welfare results in terms of employment can be shown as an important argument in terms of public consensus-building.

Discount Rates for Environmental and Transport Policies

The present debate reveals positions which are far apart and conflicting: on one side, the elaboration on the real behaviour of firms in investing (Dixit and Pindyck, 1994) points to very high 'implicit' discount rates ('option values'). On the other side, the pressure of severe environmental problems points in exactly the opposite direction.

If 'option-value-related' rates of discount are used, this implies that special attention should be paid to flexible policies, i.e. taxation, subsidies, tariffs, etc. on 'fixed' infrastructure.

This may be a good thing also for environment-related issues since large infrastructures are in general not environment-friendly. As high discount rates make programmes yielding vast gains in the far distant future (e.g. to reduce global warming) virtually worthless, recent work on environmental issues is making some economists reconsider the very purpose of discounting.[11] Alternative approaches are emerging, in a range that includes:

- use of a high discount rate for the first 30 or so years and then lower rate(s) for more distant years;
- use a similar weight on the welfare of future generations as on the welfare of today's i.e. to use a much lower discount rate (one per cent) than the one appropriate for short-term projects.[12]

One possible way to deduce society's preferences on intergenerational distribution and discount rates is *mock referenda* (Kopp and Portney, 1999), which may be organised to ask a sample/many samples of people what they think of a specific policy proposal, described in terms of cost and benefits and their expected distribution over time and space. In fact this looks a 'not-so-new' approach, its most severe weakness being the 'free riding' problem: answering to a referendum costs nothing, and so showing altruistic values seems far more easy than if actual taxes have to be paid. Nevertheless, recent developments in stated preferences surveys (that are widely used in the analysis of transport demand and whose reliability is increasingly recognised in comparison to 'revealed preferences') may help in handling such problems in the context of mock referenda.[13]

The conflict sumarised above seems to be a good example of the urgent need to elaborate explicit 'shift values' also for discount rates, in order again to limit the arbitrariness of political decisions, and to improve the overall consistency and transparency of the decision-making process.

Prices above Marginal Costs

By definition, scarcity prices are above marginal production costs, that sometimes are close to zero. Land prices are generally of such nature, and are strongly linked with accessibility, i.e. with transport. Traditional cost-benefit analysis assumes perfect competition in every sector, including land-use, i.e. no scarcity for this factor. This is a highly unrealistic assumption, that the most recent literature suggests to dismiss.[14]

In *practical* terms, cost-benefit analysis has to take in first place into account all the land value variations linked to generated traffic. Land value variations linked with existing traffic, when transport cost is reduced, are only welfare transfers, i.e. a zero-sum game. But the land-value increase proportional to generated traffic is a Pareto welfare gain (no one is losing, and land owners gain). This first approximation assumes that land values elsewhere remain unaffected, i.e. that the effects of the project under evaluation are small compared with the total land market.

A second approximation, possible if a transport-land use model is used, requires the evaluation of all the main land price variations consequent to a transport project, and the dead-weight welfare gains generated by the reduction of accessible land scarcity (the apparent paradox of *local* land value increase and the fact that land scarcity, and therefore its price, in *total* must decrease, is explained again by the 'transfer' nature of the latter price variation).

More complex is the evaluation of the effects of land prices on local monopolies. Theoretically, this is only the transfer of land benefits (i.e. accessibility) to productive or commercial activities. In practice, since firms are not perfectly mobile, and quite often land property is coincident with the ownership of the related economic activities, the overall result is extremely difficult to unbundle. Empirically, the only evidence is that lower transport costs tend to reduce the 'market power' of monopolistic and oligopolistic firms. Further work is needed on this aspect, since the consequences of these observations are relevant: reducing transport costs has larger benefits that those assumed in standard cost-benefit analysis. Therefore, for example, the welfare cost of internalising environment-related externalities may well be higher than expected.

Finally, even if not directly related to transport, the social cost of land-use constraints may also be higher than generally assumed, given the effect of reinforcing local monopolistic structures.

Conclusive Remarks

If the 'capture' of policy makers by vested interests is acceptable as a (main) explanation behind the picture presented previously, the need for evaluation tools based as far as possible on 'standard' welfare measures seems compelling in order to minimise discretionary attitudes by the decision-makers. Progress in CBA, including the developments discussed above, can therefore be expected to play an important role in improving the decision process and the quality of decision-making within the transport sector.

This emphasis on CBA does not mean that the 'standard' welfare tools are socially neutral: they ignore distribution issues, above all. But as net variations in welfare are estimated by summing up gains and losses of *all* affected subjects (from individuals to governments), they tend to make transparent the vested interests that are the main source of policy distortions and inconsistencies.

A 'non-benevolent prince' has probably to become the *standard scientific assumption*, even if this may create a problem of professional consensus among the technicians.[15] This assumption is also consistent with the creation of independent 'Transport Authorities', in charge of pure efficiency objectives, limiting also in this way as far as possible the discretionary sphere of action of the policy-makers.

Notes

1 Buchanan and Olsen (see Buchanan, 1969; Buchanan, James and Lee, 1989), only to quote two of most eminent authors in public-choice theory, are in fact consistently recommending to reduce government as much as possible and leave markets work; but even within quite different political choices and values, having a realistic picture of the behaviour and the motivations of the decision-makers seems important.

2 The ASTRA model was developed and tested in the 4th Framework Programme for R&D in Transport (EC DG-TREN) to support dynamic estimates of indirect impacts from transport polices, including on public finance.

3 Patterns in cartelization can be deducted from the actual behaviour of the involved actors: until now, incumbent rail companies were not challenged by meaningful 'aggression' in their domestic markets, even if differences in production costs at different railway companies (as reported also by UIC, Union Internationale des Chemin de Fer) would point out large potential advantages in doing so. In the absence of a meaningful political support to rail liberalisation from Member States, an example of the attitude of incumbent European rail operators toward liberalisation can be found in the refusal of the Union of the European Railways to even discuss with the authors the document on Directive 440/91 prepared on their behalf by seven academics whose conclusions, while underlying drawbacks, weak spots, undefined issues, were basically favourable toward the Directive.

4 As stated in *The Single Market Review 1996, Impact on services: Air transport* 'the most important barrier remaining (to competition) is access to airport slots, which inhibits entry from both new airlines and those that already operate some services from an airport. In accepting grandfather rights, the regulation on slot allocation has done little to alleviate this. This is currently being addressed by the Commission but it is difficult to see a solution that does not involve either confiscation or some form of auction'.

5 COM(2000)7 – 2000/02/12 (COD).

6 The estimate is extracted from an article published in *Economia Pubblica* No. 5, 2000, by Ponti, M., 'I costi esterni del trasporto e le linee politiche che ne derivano'.

7 Current estimates will raise this value up to 30 per cent: most advanced cars already reach such performance.

8 Surplus losses generated by subsidies have been estimated 'on the back of an envelope' (Petretto, mimeo 1997) equal to 0,4 euro per any euro spent by the public sector in Italy. Using a more complex model for the Italian economy (Fiorito, 2000), the author estimated surplus losses generated by subsidies equal to 0,136 Euro.

9 Topics discussed are in-depth exploration in the PROPOLIS consortium awarded by DGXII in the V framework for R&D (sustainable cities).

10 Calculations made with the collaboration of Simone Bosetti, TRT Trasporti e Territorio Srl.
11 For an exhaustive review see Portney and Weyant (editors), 1999.
12 In theory this implies compensation for delayed consumption ('utility rate of discount' or 'pure impatience').
13 Indeed, naive applications including trivial questions like 'how much importance do you attach to environmental resources?' are likely to provoke strategic responses. However, this risk has been early recognised by researchers (see, for instance, Mitchell and Carson, 1989). A careful definition of hypothetical scenarios where preferences or attitudes are linked to taxation or pricing schemes affecting the current expenditure is proved to be effective in reducing the overestimation of willingness to pay for public goods.
14 An exhaustive review was completed on behalf of the British Department of the Environment, Transport and the Regions (DETR): *Transport and the Economy.* The Standing Advisory Committee on Trunk Road Assessment, HMSO, August 1999.
15 The prince can well be benevolent *toward the technicians*.

References

Buchanan, J.M. (1969), *Cost and Choice: An Enquiry in Economic Theory*, Markham, Chicago.

Buchanan, J.M., James, M. and Lee D.R. (1989), 'Cartels Coalitions and Constitutional Policy', *Public Choice Studies*, Vol. 13.

CEC (1995), *Towards fair and efficient Pricing in Transport – Policy Options for Internalising the External Costs of Transports in the European Union*, Green Paper, COM(95) 691, European Commission, Brussels.

CEC (1998), *Fair Payment for Infrastructure Use: A phased Approach to a Common Transport Infrastructure Charging Framework in the EU*, White Paper COM(98) 466, European Commission, Brussels.

Dixit, K. and Pindyck, R.S. (1994), *Investment under Uncertainty*, Princeton University Press, Princeton, N.Y.

Fiorito, R. (2000), 'Government Debt, Taxes and Growth', *Journal of Public Finance and Public Choice*, Nr. 2/3.

Kopp, J. and Portney, P.R. (1999), 'Mock Referenda for Intergenerational Decisionmaking', in Portney and Weyant (eds), *Discounting and Intergenerational Equity*, Resources for the Future, Washington D.C.

Layard, R. and Glaister, S. (1994), *Cost-Benefit Analysis* (second edition), Cambridge University Press.

Little, I.M.D. and Mirrlees, J.A. (1974), *Project Appraisal and Planning for Developing Countries*, Heinemann Educational Books, London.

Mitchell, R.C. and Carson, R.T. (1989), *Using Surveys to Value Public Goods: The Contingent Valuation Method*, Resources for The Future, Washington D.C.

Pelkmans, J. and Labory, S. (2000), 'European Regulation and Cost-Benefit Analysis: a Methodological Guide for Non Specialists', Report CEPS – Centre for European Policy Studies, Brussels.

Portney, P. and Weyant, J. (eds), (1999), *Discounting and Intergenerational Equity*, RFF, Washington D.C.

Schleifer, A. and Vishny, R. (1999), *The Grabbing Hands: Government Pathologies and their Cures*, Harvard University Press, Cambridge, MA.

Chapter 9

Conceptual Foundations of Cost-benefit Analysis: A Minimalist Account[1]

Robert Sugden

This paper is a first attempt to set down a line of reasoning on which I have been working for some time, but which is still only partially developed. It is a response to a growing sense of unease about the validity of what are often taken to be the fundamental assumptions of cost-benefit analysis (CBA) – an unease which, for me, coexists with a conviction that CBA is a genuinely useful technique for policy analysis. My aim is to free CBA from certain assumptions that are vulnerable to criticism, by showing that these assumptions are not necessary. I hope that this preliminary account will promote debate about the foundations of CBA.

I have chosen the title of the chapter with some care. I am not undertaking to set out *the* conceptual foundations of CBA. CBA is a body of practical techniques. Although it is clear enough that these techniques are largely derived from neo-classical economic theory, what we take the conceptual foundations of CBA to be is still to some extent a matter of preference and judgement. My aim is to offer *a* coherent account of those foundations, and to explore its implications.

Notice too that I am writing about the *conceptual* foundations of CBA, not the *welfare* foundations. It is often said that CBA is a form of applied welfare economics, and in a loose sense that is clearly true. However, on a strict interpretation, welfare economics is the study of how welfare or well-being is affected by economic variables. On the account I offer in this paper, whether CBA measures well-being is an open question, external to CBA itself. Nor am I writing about the *normative* foundations of CBA. Some welfare economists define their sub-discipline as the study of the economic implications of given normative principles. From this perspective, CBA might be interpreted as a practical form of normative enquiry, based on particular value judgements. But on the account I offer, CBA does not purport to tell us anything directly about what is good or bad, or about what ought or ought not to be done. Rather, it provides useful *information*, in a compact and standardised form, to all those who are involved in the relevant decision-making process.

In calling my account *minimalist*, I signal my intention that the foundations I construct should depend on as few assumptions as possible. In particular, I hope to de-couple CBA from two assumptions that underlie much of neo-classical economics, but that have been subject to a great deal of recent criticism. The first is the *consistency assumption*: that individuals have stable preferences, and that those preferences satisfy the strong principles of completeness and internal consistency that are postulated in conventional decision theory. The second is the *revealed welfare assumption*: that an individual's preferences provide a measure of her well-being, or of what is good for her.

The consistency assumption is empirical, and therefore open to test. In the last two decades, there have been many controlled tests of conventional decision theory, and the results have generally been discouraging. Various predictable patterns of behaviour, incompatible with the conventional theory, have been found to be characteristic of most human decision-makers. Some of these patterns cause serious problems for CBA. For example, the phenomenon of *preference reversal* suggests that the mental routines that people use when choosing between two options are different from those that they use when putting a money value on a single option. The now well-known disparity between willingness-to-pay (WTP) and willingness-to-accept (WTA) valuations – a disparity that is far too large to be explained by the income and substitution effects of conventional theory, and which is observed even when those effects are screened out by experimental controls – suggests that people's preferences are conditional on the *reference points* from which they view decisions. The problems caused for CBA by such deviations from conventional theory have become more obvious with the growth of stated preference and contingent valuation methods, which try to infer preferences from survey data; but it is becoming clear that the deviations are not artefacts of survey methods: it is the theory, not the surveys, that is at fault.

In some cases, deviations between theory and survey data seem to be the product of certain kinds of normative attitude on the part of individuals, which may not be compatible with the standard model of preferences. For example: in the conventional theory, each individual has private preferences over public goods, with the implication that WTP valuations for a public good can be elicited independently from each individual and then summed. But experimental and survey evidence suggests that many people's willingness to contribute to public goods is governed by ideas of fairness and reciprocity: an individual is willing to contribute what she sees as a fair share of the costs of a public good, if and only if other people contribute fairly too. Notice that such notions of fairness are not preferences about final outcomes. (Person A's desire that person B contributes fairly is not a preference on the part of A about the level of B's final consumption: it is an attitude to the *procedure* by which final outcomes are reached.) It is not obvious that such non-consequentialist normative attitudes can coherently be represented in terms of individuals' preferences, as conventionally understood.[2]

The empirical failure of the consistency assumption causes problems for anyone who wants to justify the revealed welfare assumption. There are strong reasons for expecting defensible judgements about individual well-being to satisfy certain consistency conditions. As a matter of empirical fact, a person's *preferences* may be unstable and context-dependent; as a matter of normative principle, we may respect the choices that result from those preferences (in the sense of not wanting to overturn them); but it seems wrong to say that the nature of the person's *well-being* is unstable or context-dependent.[3] If judgements about well-being are required to satisfy consistency conditions which are not in fact satisfied by preferences, the revealed welfare assumption cannot be sustained.

The revealed welfare assumption has also come under attack from a different direction, from philosophers and philosophical economists who have set out to analyse the nature of well-being (or – for those who start from a different meta-ethical position – the nature of widely-held judgements about well-being). The work of these theorists casts doubt on the credibility or persuasiveness of the claim that a person's preferences, even if internally consistent, reliably reveal what is good for her.[4]

Since my minimalist approach makes no normative claims about well-being, it clearly insulates CBA against criticisms of the revealed welfare assumption. I cannot claim that it resolves all the problems caused for CBA by the failure of the consistency assumption; but I suggest that it is worth considering as a possible starting point for tackling those problems.

CBA as the Measurement of Economic Surplus[5]

In CBA, as I characterise it, benefits and costs are defined in terms of *individuals' willingness to accept one thing in exchange for another.* An individual receives a benefit whenever he receives something in return for which he is willing to give up something else that he values. To measure how large that benefit is, we measure how much he is willing to give up to get it. Conversely, an individual incurs a cost whenever she gives up something that she would willingly give up only if, as compensation, she was given something else that she valued. To measure how large that cost is, we measure how much she would accept as compensation for incurring it.

These formulations define benefits and costs in terms of one another. This is not a circularity: it reflects the fact that, within the theoretical framework we are using, value is *relative.* The scale on which, for any given individual, value is being measured need not have any interpretation beyond 'exchangeability'. In particular, we do not need to claim that it is *better* for the individual that she has those things that she values more highly, or indeed that her having those things contributes to any objective at all (psychological satisfaction, happiness, well-being, or whatever). Perhaps the individual consistently acts on preferences which rank outcomes according to some standard that she takes as her objective. If so, our measurements of costs and benefits to her will use that standard; but we do not have to endorse her objective as good for her. But perhaps she simply has no consistent objective. That does not mean that we cannot measure the relative valuations she reveals in particular contexts.

This approach allows all costs and benefits to be measured in a single dimension if, *as a matter of convention,* we choose one particular type of benefit to use as a standard. We can then express all other benefits and costs in terms of that standard, using individuals' own preferences to determine equivalences of value. If we are to use the same standard of measurement for all individuals, the standard has to be a finely divisible good that everyone prefers to have more of rather than less, and that individuals treat as a potential substitute for the array of benefits and costs that we want to measure. ('Substitute' is here used in a subjective sense: with respect to the preferences of a given individual, two goods are substitutes for one another to the extent that that individual is willing to accept a gain of one as compensation for a loss of the other.) The usual convention is to use *money* as the standard of measurement.

If money is used as the standard, the measure of benefit is WTP. That is, a benefit to any given person is measured by the maximum amount of money that that person would be willing to pay in return for receiving the benefit. Similarly, the measure of cost is WTA. That is, a cost to any person is measured by the minimum amount of money that that person would be willing to accept as compensation for incurring the cost. *Economic surplus* is generated in a transaction if a person receives a benefit and pays less than her WTP, or if she incurs a cost and receives more than her WTA in compensation; surplus is defined as the excess of WTP over actual payment, or of actual compensation over WTA.

These measures of cost, benefit and surplus clearly presuppose some degree of consistency of preferences. Specifically, they presuppose that, *in any given context*, any individual prefers to pay less money rather than more for a benefit, and prefers to receive more money rather than less as compensation for incurring a cost. But there seems little reason to expect these assumptions to be far wrong. Thus, provided the context of measurement is well-defined, these measures do indeed rest on minimalist assumptions. When we are considering actual transactions in markets, based on mutually-agreed terms which involve the exchange of money for goods or services, the relevant context is defined by the transaction itself. For example, suppose a person makes a particular trip by train on a particular day, paying some particular fare. To measure the surplus he derives from the opportunity to travel by train, we have to find the highest fare that he, on that day, would have been willing to pay for that trip. We do not have to assume any more consistency in his preferences than this: that for any two sums of money x and y, if he is willing to pay x, and if y is less than x, then he is willing to pay y. Real difficulties start to arise only when the context of measurement – the setting in which notional exchanges are made – is not well-defined. I shall return to this problem later.

Since surplus accruing to different individuals is measured in the same money units, it is meaningful to sum surplus across individuals. For example, suppose persons A and B are parties to some transaction, in which A buys some object from B. Suppose A's WTP is £15, B's WTA is £8, and they settle on a price of £10. Then the transaction generates £5 surplus for A and £2 for B. There is nothing problematic in summing these surpluses, and saying that the transaction generates £7 surplus for A and B together. Or suppose that A engages in an activity which imposes external costs on B. Suppose, as before, that A's WTP is £15 and B's WTA is £8, but suppose that B's consent is not required and thus no compensation is actually paid. Then the activity generates £15 surplus for A and –£8 for B. Again, there is nothing problematic in saying that the activity generates £7 net surplus for A and B together. All this may seem obvious; but it is important to recognise that these summations do not involve interpersonal comparisons of well-being.[6] In adding A's surplus to B's, we do not have to claim that each £1 of surplus accruing to A is just as good for A as each £1 of surplus is good for B, or that A's £1 and B's £1 are just as good as one another from some social point of view. All we have to say is that £1 to A and £1 to B sum to £2 to A and B together. The idea of aggregating *well-being* across persons raises deep methodological and philosophical problems; but aggregating *surplus* is just a matter of adding amounts of money.[7]

CBA, as I characterise it, measures the economic surplus generated by a *project* (by which I mean any specific proposal which gives rise to costs and benefits). The bottom line of a CBA is the *total* economic surplus generated, i.e. the net sum of the surpluses generated for all individuals. Disaggregations of this total, which identify the net surplus accruing to particular groups of individuals, may be provided as additional information.

I have stressed that CBA is an exercise in measurement: it does not prescribe what decisions ought to be taken. However, it would be disingenuous to deny that I am appealing to the idea that the creation of surplus is, in some general sense, a good thing. Someone who has no sympathy with this idea is unlikely to find CBA useful or informative. So what is good about the creation of surplus? Economic surplus, I suggest, is what we mean (or have to mean) by 'wealth' when we talk about market processes 'creating wealth'.

At the most fundamental level, much of the driving force of a market economy comes from attempts to seek and to appropriate surplus. Firms are motivated to produce goods and services that consumers want to buy, to produce these at the lowest possible cost, and to sell them at the highest possible price. By buying cheap and selling dear, firms appropriate surplus. This does not have to be true of all firms; what matters is that there are many firms that are motivated in this way, so that opportunities to appropriate surplus are actively sought out and exploited. In exploiting these opportunities, each firm is constrained by the activities of its competitors and by consumers' and workers' opposing attempts to sell dear and to buy cheap, that is, to appropriate surplus for themselves. A competitive market, then, is a spontaneous order resulting from the interaction of surplus-seeking agents.

Under conditions of perfect competition, any activity by one person or firm that generates surplus for that person or firm also generates positive net surplus for everyone taken together. This is not because the activity has no significant effect on other people, but because its effects on other people, measured in terms of surplus, cancel out.[8] For example, suppose that person A decides to go into business as a self-employed taxi driver. The surplus that this project generates for A is the difference between his WTA for the time he foregoes and the net revenue of his business (i.e. revenue from sales minus expenditure on inputs other than his own labour). In a competitive taxi market, A's entry will reduce the price of taxi services. This price reduction will generate positive surplus for taxi users and negative surplus for other taxi drivers. Although the price change may be small, it affects many people; the gross changes in surplus, summed over all taxi users or summed over all other taxi drivers, are of the same order of magnitude as the revenue of A's business. However, these two changes in surplus cancel out. That they cancel out does not depend on any assumptions about preferences other than that, in relation to any given transaction, taxi users prefer to pay less rather than more, and taxi drivers prefer to receive more rather than less.

To the extent that real markets approximate to the competitive model, normal market transactions generate positive net surplus for everyone taken together. Similarly, projects carried out by firms generate positive net surplus if and only if they are privately profitable. Conversely, if opportunities exist to generate positive net surplus, the competitive environment ensures that individual agents can appropriate surplus to themselves by exploiting them. Thus, in a competitive market, it is *as if* each agent is seeking out opportunities to generate positive net surplus for everyone taken together. For this reason, competitive markets are highly effective at creating surplus.[9]

The measurement of net surplus through CBA, and the use of the *net surplus criterion* (or *efficiency criterion* or *compensation test*) which approves projects if and only if they generate positive net surplus, can be interpreted as a generalisation of the decision-making procedures of competitive markets. Of course, this interpretation does not *justify* the net surplus criterion, either for private decision-making in markets or for decisions about the supply of public goods. But it does help to *familiarise* the criterion: it helps us to understand what it means, and to assimilate it to other strands of economic and political thought. We can say that CBA, if based on this criterion, is an attempt to *simulate* the workings of competitive markets where they do not exist.

It is a fact of current economic life that the allocation of resources among private consumption goods *is* generally determined through markets, and that recent trends in social and political thought have been towards greater rather than less reliance on markets. Thus, it is difficult to imagine informed public support for any method of evaluating projects which does not endorse the allocations of resources that are generated by the normal operation of markets, in the absence of obvious sources of market failure. That the net surplus criterion does endorse such allocations is a strong point in its favour – in the practical sense that people are likely to find the criterion useful and informative.

CBA as an Aid to Informed Debate

Another very obvious current trend in current political thinking is towards greater openness and accountability in public decision-making, and away from practices which depend on deference to professional and political élites. This trend is hostile to some established interpretations of CBA. At least in the Anglo-Saxon world,[10] there has been a long-standing tradition in which governments have seen themselves, and have been seen by their policy advisers, as benevolent maximisers of social welfare. In this perspective, CBA and other techniques of policy guidance have been seen as ways of giving advice to governments about how to make decisions in the best interests of society as a whole. There has been little recognition of any political process that involves actors other than government decision-makers (construed as ministers of state, or as senior civil servants). Thus, the idea that CBA might inform public debate, or help the public to hold governments accountable for their decisions, has been overlooked. As an icon of this approach, I offer the following remark from a manual of CBA written in the seventies. Arguing in favour of weighting benefits differently according to the income of the beneficiary, the authors suggest that it might be a good idea to keep this practice secret: 'It may sometimes be politically expedient to do good by stealth' (Little and Mirrlees, 1974, p.55). I doubt whether any economist would express such a sentiment so openly today; but the current practice of CBA still reveals traces of its utilitarian history.

The traditional idea that CBA is addressed to a unitary 'government', construed as if it were a benevolent despot, also fails to recognise the fact that more than one level of government – local, regional, national, supranational – can be involved in a decision about a given project. Further, projects may be joint ventures between government agencies and private firms; and (particularly when projects are financed by international loans) some of the residual risks may be borne by banks. In all these cases, decision-making is a matter of negotiation. CBA is not used simply to help one party – government agency, firm or bank – to decide on its negotiation position; it is also used within the process of negotiation, for example, when one firm, agency or tier of government sponsors a project and seeks co-funding from another. If CBA is to be useful in such negotiations, openness and public credibility are essential.

I suggest that CBA is best understood as a means of informing discussions about the merits and demerits of projects. It performs this function by supplying a particular kind of information – information about economic surplus – in a generally-understood format.

As an analogy, consider the way in which financial accounting data can inform decision-making in organisations which pursue non-profit objectives in a market environment. For example, consider the governing board of an art gallery deciding whether to upgrade its cafeteria. The board would presumably find it useful to see financial projections of the costs and revenues of this project, even though financial considerations are not the only ones to be taken into account. (The ultimate objective, I take it, is to have an excellent art collection which gives pleasure and education to a large number of visitors. A good cafeteria attracts more visitors and makes the visitors' experience more pleasurable. However, finances matter too: losses on cafeteria operations detract from other ways of pursuing the gallery's objective, such as buying new works of art.) The meaningfulness of financial accounting data derives from the existence of reasonably well-defined accounting conventions. Because these conventions are constant across accounting exercises carried out at different times, for different purposes and in different organisations, the members of the board are likely to have a good sense of what the numbers in any one such exercise mean. For the same reasons, outsiders are able to interpret the financial evidence that the board uses, and to judge for themselves what decision the board should take. In addition, the existence of well-established accounting conventions provides a check on the competence of the accountants who have made the projections. *Ex ante*, the plausibility of the projections can be checked by comparing them with projections made for, and results achieved by, similar projects. *Ex post*, the projections can be compared with the actual results.

CBA, interpreted as a measure of economic surplus, takes account of a wider range of costs and benefits than financial accounts do. But the definition of 'cost' and 'benefit' is still tight; the way in which costs and benefits are to be valued is governed by clear conventions. Most importantly, relative valuations in CBA are taken from the preferences of those individuals who receive the benefits or incur the costs; they are not inserted by the analyst or the decision-maker. In this respect, CBA differs sharply from multi-criteria analysis, which allows each decision-making context to have its own set of criteria, and which requires an exogenous system of relative weights, specific to that context, for those criteria. Because the conventions of CBA are tightly-defined, the results of different CBA studies are comparable across types of decision, across government agencies, across time and across countries.

The scope for such comparisons provides a valuable check on the credibility of individual studies, tending to screen out both inadvertent errors and the deliberate use of *ad hoc* assumptions to bolster the case for a particular project or to rationalise particular decisions. It also allows members of the general public to build up a body of experience which allows them to interpret the results of CBA studies for themselves, and to make informed contributions to political debates in which those results are used as evidence. Of course, it would be unrealistic to expect most ordinary citizens to have a *direct* understanding of CBA, any more than they understand the conventions of financial accounting. But in a civil society, many organisations and occupations (pressure groups, political parties, consumer associations, academics, journalists, broadcasters, and so on) have specialist skills in examining and criticising government decisions on behalf of citizens in general. If common conventions of CBA were in general use, we might expect those conventions to be understood – at least in a rough and ready way – at this level.

It is an inescapable implication of the approach I am advocating that CBA does not necessarily take account of all factors that impinge on public debate, or that public decision-makers may quite properly want to take into account. But this should not be interpreted as a *failure* of CBA, any more than in the case of the art gallery it is a failure of financial accounting that it does not take account of the objectives of artistic excellence and education. The result of a CBA is just one piece of information, to be considered alongside others. Nor should we *necessarily* conclude that we need some higher-level theoretical framework, such as multi-criteria analysis, within which economic surplus counts as one criterion among others. In an open and democratic society, no technical apparatus of decision support can replace the political process. Ultimately, decision support has to be a matter of making relevant information available to decision-makers and to their legitimate critics, the general public. My claim is that information about economic surplus is useful in its own right, independently of any information that might be supplied by multi-criteria analysis.

In the appraisal of transport projects, two kinds of project impact stand out as featuring strongly in public debate about transport, while not explicitly being registered in CBA. The first is *economic development*. Many transport projects are advocated on the grounds that they promote the economic development of particular areas. This argument is particularly used in support of improved links between economically depressed 'peripheral' areas and more rapidly-growing 'core' areas. A variant of the argument is often used in support of urban public transport projects, particularly rapid transit systems, as a means of promoting and sustaining economic growth in large cities. The second impact is *social inclusion*. The expansion of car-ownership, the associated decline of inner-city employment, local shopping centres and public transport, and the perceived increase in street crime are seen as creating classes of deprived people (the young, the old, the handicapped, single parents of young children) who lack access to the opportunities created by economic growth. Public transport and traffic-calming projects are proposed as means of providing such access.

Advocates of CBA can reply that CBA *does* take account of economic development and social inclusion, and that any attempt to treat these factors as additional impacts of transport projects would be double-counting. At a theoretical level, this reply is substantially correct.[11] It is correct to the extent that transport projects contribute to economic development and to social inclusion only *by means of* reducing the real cost of trips. As an extreme example, imagine that a toll bridge is built to link an island to the mainland, but that the tolls are set so high that no one uses it. Then the bridge would have no economic development impacts. If the bridge has economic development impacts (positive or, conceivably, negative), these occur as adjustments to the reduction in the cost of travel that it brings about. Similarly, if wheelchair-accessible buses were never ridden on by wheelchair users, they would not contribute to the social inclusion of wheelchair users; if footpaths and cycle paths were never used, they would not contribute to the social inclusion of people without cars. A comprehensive CBA would predict all changes in the costs of trips (including trips made on foot) resulting from a project, and all induced changes in travel patterns; it would then measure the economic surplus created by those changes. If the object of CBA is to measure changes in economic surplus, nothing more needs to be done.

So it is not a legitimate objection that CBA, as conventionally applied to transport projects, *ignores* their impacts on economic development or on social inclusion. What it perhaps fails to do effectively is to *provide specific information* about those impacts – information that is relevant for people who, in judging the merits of projects, attach particular importance to those impacts. For example, in a CBA of a road project, the benefits arising through reduced travel costs can be estimated by using only projections of trip costs and trip volumes, with and without the project. This may not be the most relevant information for predicting development impacts. (For instance, it might be more important to predict changes in the geographical distribution of shopping *expenditure* than to predict changes in the number of shopping *trips*; but only the latter is necessary for a CBA.) Thus, effective decision support may require the development of summary statistics for describing the impacts of projects on economic development or on social inclusion, which are parallel to the information provided by CBA. But this is not a reason to revise the principle that CBA measures economic surplus, and only economic surplus.

Payment Mechanisms

All measurements of economic surplus depend on counterfactual propositions about amounts of money that individuals would or would not pay, or would or would not accept, in hypothetical situations. Even in the simplest case of surplus associated with private consumption goods bought in markets, the calculation of surplus depends on propositions about how much consumers would buy at hypothetical prices. In such cases, however, cross-section and time-series investigations of demand can usually give reasonably reliable predictions about responses to price changes (at least within the range for which there is experience). Things become more difficult when the surplus we are trying to measure is associated with goods that are not directly traded in markets, so that propositions about WTP and WTA have a much greater hypothetical content.

In order to measure surplus on non-marketed goods, we have to define a *payment mechanism* – a hypothetical process by which individuals are required to pay for benefits they receive, or by which they are free to refuse to allow costs to be imposed on them unless they are compensated. WTP or WTA *within that mechanism* then provides the measure of economic surplus. For example, suppose we want to measure the surplus generated by a project which improves a road network to which there is free access. The notional payment mechanism could be a toll levied per trip, or an access charge per day, or an access charge per year, or any of many other possibilities. In each case, we would try to discover the highest payment each individual would be willing to make, through the relevant payment mechanism, for the benefits generated by the project.

In the absence of strong assumptions about attitudes to risk, about the amount of information that individuals have, and about the internal consistency of individuals' preferences, it is perfectly conceivable that the economic surplus generated by a given project differs according to the payment mechanism used. Consider an analogy in the domain of marketed goods. Suppliers of consumer durables often offer service contracts to buyers: in return for a fixed annual fee, the supplier undertakes to repair the appliance if it breaks down. By estimating the demand function for such contracts, we can measure the economic surplus generated by the repair service. But exactly the same repair service could be offered on a pay-as-you-go basis. By estimating the demand function for individual repair episodes, and by predicting the rate of such episodes for each appliance, we could come up with another measure of the surplus generated by the service. Would the two measures be equal? Not necessarily. One possibility is that consumers are risk-averse, and are willing to pay a premium for the element of insurance provided by a service contract. Another is that they are ill-informed. For example, if they tend to over-estimate the probability of breakdowns, surplus measured with respect to the price of service contracts will be greater than surplus measured with respect to pay-as-you-go contracts. A third possibility is that their preferences do not satisfy conventional assumptions about internal consistency. For example, if people do not discount future costs and benefits at a constant rate, but always give disproportionate weight to those that occur in the present, they may be averse to contracts for which they have to pay in advance; in consequence, surplus may be greater for pay-as-you-go contracts.

If measures of surplus differ according to the payment mechanism used, that does not invalidate the concept of economic surplus: it is simply a fact about the economic world. I have argued that the driving force of a market economy comes from attempts to seek and to appropriate surplus. The same wealth-seeking, competitive dynamic that leads firms to minimise production costs and to give their products the characteristics that consumers are willing to pay for also induces them to seek out the most profitable payment mechanisms for selling those products. In a perspective in which surplus, rather than preference, is the fundamental concept, it is artificial to try to distinguish between these different ways of creating surplus. We have to recognise that surplus is specific not just to particular goods, but also to particular payment mechanisms.

However, this property of the economic world causes severe difficulties for CBA. The problem is that, in CBA, payment mechanisms are often merely hypothetical. Take the case of a typical road project, which will be free at the point of use. If it goes ahead, individuals will in fact pay for it, as citizens, through taxation. But taxation is not a payment mechanism in the sense I am using the term, i.e. a mechanism through which individuals decide, as individuals, whether to pay money to receive benefits, and in so doing, reveal their WTP. There simply is no 'true' payment mechanism in this case. In order to define a concept of surplus for a project of this kind, we have to stipulate a *hypothetical* payment mechanism.

In proposing conceptual foundations for CBA, I have deliberately avoided any presupposition that each individual has a well-defined set of 'true' preferences. But having taken this line, I cannot propose, as a criterion for selecting a hypothetical payment mechanism, that we look for the mechanism that would be most likely to reveal individuals' true preferences. So what criterion can we use?

I think we must accept that the ultimate status of any such criterion will be that of a convention (in the same sense that many accounting practices are conventions). But if we accept the idea that CBA is an attempt to simulate the workings of competitive markets, it may be possible to defend some broad principles for choosing among hypothetical payment mechanisms. I now suggest two such principles.

The first is the principle of *privateness*: if possible, the hypothetical payment mechanism should be a market-like environment, in which people, acting individually, buy or sell private goods. Thus, in measuring the surplus generated by a public good, we should try to find payment mechanisms which involve private correlates of that public good. I take this principle to be a corollary of the idea that CBA is an attempt to simulate markets.

For example, consider again the case of an improvement to a road network to which there is free access. The network itself can be considered as a public good: if it is supplied to one potential road user, it is supplied to all. However, *trips* on the network are private goods (private to the individuals who make them) which just happen to be priced at zero. In selecting a hypothetical payment mechanism, we should choose a mechanism such as tolls, in which individuals pay for private goods, rather than a mechanism in which individual citizens hypothetically dictate public decisions between alternative packages of road networks and tax burdens. Similarly, if we want to measure the surplus that people derive from using a recreational area, we should focus on the private good of individual access to the area, rather than on the public good of the area's existence.[12]

The second principle is that of *pay-as-you-go*: we should favour hypothetical payment mechanisms which allow individuals to reveal their WTP and WTA for benefits and costs at the time they occur, and in the state of the world in which they occur. (This principle is subject to some qualifications, which I shall explain later.) For example, suppose that now (the year 2001) we are measuring the benefits that will be generated by a rapid transit system which will come into operation in some city in 2006. One approach would be to assume that the only way anyone can use the system is for her to buy, in 2001, a ticket giving her free use of the system for life. We could then consider the set of all people who might be residents of the city in 2006, and try to discover the sum of all such people's WTP *now* for such a ticket. An alternative approach would be to assume that people who use the system will pay separately for each trip. We could then try to predict, for each day of the system's operation, the total WTP of users *on that day*. The second approach implements the pay-as-you-go principle; the first does not.

If we use the first approach, the measurement of surplus reflects *individuals' own* predictions about their future preferences and about the characteristics of the project, and their own judgements about the relative probabilities of different states of the world. In contrast, if we use the pay-as-you-go approach, the measurement of surplus reflects the *cost-benefit analyst's* predictions and judgements. Remember that CBA does not need to measure the surplus accruing to particular individuals; the aim is merely to measure *total* surplus. Many of the person-specific factors that create uncertainty at the level of the individual (for example, each individual's uncertainty about where he will live and work in future periods) cancel out at the aggregate level because of statistical independence and the law of large numbers. Thus, it may be easier to make reliable predictions at the aggregate level than at the individual level. In any case, I shall assume that the people to whom CBA is addressed – decision-makers and the general public – take the analyst's predictions and judgements to be reliable. Given this assumption, what is the justification for using pay-as-you-go payment mechanisms in CBA?

Again, consider the analogy between CBA and competitive markets. For consumer goods and services, pay-as-you-go pricing is the norm. When a firm introduces a new product, it does not normally make advance contracts with consumers. The product is introduced *in the expectation* that when consumers see the product, they will be willing to pay the price the firm charges. Often, this expectation is grounded on specialist expertise and private information (for example, the work of market researchers). Often, too, there is an element of entrepreneurial judgement. Successful firms anticipate consumers' future preferences. Markets would be much less dynamic than in fact they are, if the criterion for the development of a new product was consumers' willingness to accept the costs in advance.

As the earlier example of service contracts for consumer durables shows, pay-as-you-go pricing is not the universal rule in markets. However, specialists in competition policy are generally suspicious of *bundling* – practices by which suppliers tie distinct products together to be sold as single packages. Such practices are often the means by which firms use a dominant position in one market to exclude competitors in another. (Thus, suppliers of consumer durables may use their privileged initial contact with buyers to gain a competitive edge in the market for repair services.) Bundling reduces the transparency of price competition, and makes it harder for consumers to make informed decisions. I suggest that, in choosing between alternative hypothetical payment mechanisms, CBA should favour those that are most transparent. This creates another presumption in favour of pay-as-you-go mechanisms.

Now for my qualification to the pay-as-you-go principle. For certain types of market transaction, the *raison d'être* is the exchange of claims contingent on different states of the world. In particular, consider the insurance industry. Does this industry create economic surplus? From an *ex ante* viewpoint, it clearly does: individuals' WTP for insurance contracts exceeds the costs of supplying those contracts. But from an *ex post* viewpoint, it does not. In any given state of the world, some policy-holders lose (they pay their premiums and receive nothing in return) while others gain (they make claims which are settled in their favour). *Ex post*, the losers' WTA for their losses is equal to the amount they pay to the insurance company; the gainers' WTP for their benefits is equal to the amount they receive; and, because of the costs of running an insurance business, the sum of the losses is greater than the sum of the gains. A similar analysis applies to the gambling industry: here, too, surplus is created *ex ante* but not *ex post*. In both these cases, the bundling into a single contract of costs and benefits contingent on different states of the world is not gratuitous: it is a service that is offered for sale on transparent terms, which consumers are willing to pay for. I suggest that CBA should evaluate such contracts from an *ex ante* viewpoint.

The treatment of accident risks in CBA provides a significant test case. In principle, the net surplus generated by changes in accident risks can be evaluated either *ex ante* or *ex post*. Suppose we are dealing with an increase in exposure to risks of injury. An *ex post* approach would project the expected increase in the number of accidents and then predict, for each person actually injured, her WTA for the relevant injury. An *ex ante* approach would measure, for each person exposed to the increased risk, her WTA for that increase in risk. The *ex ante* approach to accident risks is now the norm in CBA (although it has not always been so, and some commentators continue to argue against it).[13] It seems to me that the *ex ante* approach is indeed preferable in the case of accident risks. By the nature of accidents, there is no way of designing a payment mechanism in which individuals can choose under conditions of certainty whether or not to accept compensation in return for incurring the consequences of an accident. In this sense, the bundling implicit in *ex ante* surplus measurement is not gratuitous. Here, I suggest, the *ex ante* approach allows a closer approach to the ideal of simulating the workings of competitive markets. But I cannot conceal the fact that there is a grey area here.[14]

Preference Inconsistencies

Given the conceptual framework I am proposing, the measurement of surplus does not require the consistency assumption. Surplus is measured in relation to specific payment mechanisms, that is, in relation to specific descriptions of the context (real or hypothetical) in which individuals make decisions about whether to pay to receive a benefit, or whether to incur a cost in return for compensation. Since the consistency assumption is concerned with the relationship between the preferences that a person reveals in one decision-making context and those that she reveals in another, it does not impact directly on the measurement of surplus. Nevertheless, failures of this assumption do create a real problem for CBA, as I have characterised it.

The problem is essentially the same as the one I discussed above: the problem of making a principled (rather than arbitrary or ad hoc) choice between alternative payment mechanisms for the purposes of CBA, given that these mechanisms are merely hypothetical. The evidence from experiments and survey research suggests that minor changes in the framing of tasks (for example, the difference between 'choice' and 'valuation' framings that induces preference reversals) can lead to major differences in the preferences that people reveal. If this is so, how can we hope to defend any one framing – i.e. any one specification of the payment mechanism – as more appropriate for CBA than all the others?

I have to confess that this is a tough problem for my approach, and that I have yet to solve it at all adequately. However, I can offer some speculative comments. I conjecture that the consistent use of the privateness and pay-as-you-go principles would greatly reduce the difficulties caused for CBA by preference 'inconsistencies', i.e. deviations between the properties of preferences as elicited from surveys or from observations of behaviour, and the properties assumed in conventional theory.

In the introduction, I argued that people's willingness to contribute to public goods is often influenced by normative attitudes (for example, attitudes to fairness) that may not be capable of being represented within the conventional model of individual preference. When contingent valuation studies attempt to elicit individuals' WTP for public goods, apparent inconsistencies in people's stated preferences may reflect these limitations of the conventional model.[15] Such problems would be avoided if the hypothetical payment mechanisms used to elicit WTP satisfied the principle of privateness.

At present, I can support my conjecture about the pay-as-you-go principle only by means of examples and simple models, specific to one particular (but very significant) form of preference inconsistency: the disparity between WTP and WTA valuations that is found in so many contingent valuation exercises. This disparity seems to be the result of *loss aversion* or *status quo bias* – a fundamental asymmetry in people's attitudes to gains and losses, which makes them averse to movements away from whatever they take to be the status quo or 'reference point'. One of the most striking features of the experimental evidence is how easily people can be induced to change their reference points. A subject who comes into an experimental laboratory and is offered the opportunity to buy a box of chocolates takes her reference point to be the state in which she has no chocolates. But if the same subject is given a box of chocolates at the start of the experiment, and then a few minutes later is offered the opportunity to sell the chocolates, she takes her reference point to include her ownership of the chocolates (Kahneman, Knetsch and Thaler, 1990; Bateman *et al*, 1997). The implication is that, while people are reluctant to make exchanges which take them away from a perceived status quo, what is perceived as the status quo rapidly adjusts to new circumstances. Thus, status quo bias may be a transitory phenomenon. The fact that it has a big impact on individuals' responses to survey questions which require them to think about changes from their current status quo does not necessarily imply that it will have a similar impact on pay-as-you-go valuations.

Consider again the example of the firm introducing a new product. To be more concrete, suppose the new product is a transport link which, in comparison with current services, is faster but has a higher price (the rail link between Heathrow Airport and central London is an example). If people are subject to status quo bias, we should expect that a questionnaire survey conducted before the introduction of the service would indicate relatively low WTP on the part of potential users. But if this bias is transitory, the firm would be unwise to base its demand predictions on such estimates of WTP. If the new service comes into operation, it will quickly become part of people's perceptions of the status quo; thus, in relation to the pay-as-you-go pricing mechanisms that will be used, WTP may be greater than the survey data suggests. This is not to say that the firm should not use survey research methods, but only that, in designing survey instruments and in interpreting survey responses, it should take account of the *duration* of status quo bias. It might be more informative to use survey questions which present choices between alternative modes of travel, *neither of which* is framed as the status quo.[16]

In the Appendix, I explore this intuition in a simple formal model. The model is of a market for a private good. Preferences are reference-dependent, i.e. each individual's preferences over consumption bundles depend on the reference point from which bundles are viewed. There is status quo bias, and so there is a disparity between WTA and WTP valuations; but this bias is transitory. It turns out that the long-run equilibrium of the market is independent of the degree of status quo bias; it is determined only by the valuations of 'entrants', i.e. people confronting the market without previous experience. When the market is subjected to a shock (for example, an exogenous change in demand), short-run price changes reflect consumers' status quo bias, but in the long run, the effects of this bias disappear. The more transitory the bias is, the shorter the duration of these short-run effects.

Conclusion

This paper has been unashamedly speculative. I have offered a rough sketch of how CBA might rest on foundations that do not require us to assume either that individuals have consistent preferences, or that an individual's preferences provide a measure of her well-being. For cost-benefit analysts to make the first of these assumptions is to leave a hostage to fortune: all the indications are that the assumption is far from representing reality. To make the second assumption is to tie CBA to a particular normative position that many people do not share, and that will become increasingly difficult to defend if the evidence of preference inconsistencies continues to accumulate. I hope I have persuaded the reader that it is worth thinking about whether CBA really needs those assumptions.

Appendix: A Model of the Impact of Loss Aversion in a Market

Consider the following simple model of the market for some good G. There is a succession of discrete time periods. In each period, the number of consumers is n. (This number is taken to be large, so that the law of large numbers applies: that is, statistically expected outcomes are close approximations to actual outcomes.) At the end of each period, rn consumers *retire* from the market; at the start of the next period, an equal number of *entrants* replace the retirees of the previous period. These entrants join the $(1 - r)n$ *experienced consumers* who have survived from the previous period. In each period, retirees are a random sample of the current set of consumers.

I shall assume that G accounts for only a small proportion of consumers' total expenditure; thus, income effects can be taken to be negligible. In each period t, for each t-period entrant, the quantity of G demanded depends only on the *price* in that period. For any period t and for any price p in that period, the quantity demanded by each t-period entrant is f(p); f(.) is the *entry demand function*. The *entry valuation function* v(.) is defined as the inverse of f(.), i.e. for each quantity q, v(q) is the price p at which the quantity demanded by each entrant is q. I assume hat v(.) is continuous and strictly decreasing.

For each experienced consumer, the quantity demanded depends not only on the current price, but also on his consumption of G in the previous period. The *experienced demand function* (constant across consumers and periods) is written as g(., .); if the price in period t is p, the quantity demanded in period t by a consumer who consumed the quantity q' in period $t - 1$ is g(p, q'). The *willingness-to-accept (WTA) function* $w^-(., .)$ is defined so that for each quantity $q < q'$, $w^-(q, q')$ is the price p at which q is the quantity currently demanded by a consumer who consumed q' in the previous period. The *willingness-to-pay (WTP) function* $w^+(., .)$ is defined so that for each quantity $q > q'$, $w^+(q, q')$ is the price p at which q is the quantity currently demanded by a consumer who consumed q' in the previous period. I assume that $w^-(q, q')$ and $w^+(q, q')$ are continuous and strictly decreasing in q for all values of q'. Further, and crucially, I assume that $w^-(q, q') \geq v(q)$ for all q, q' satisfying $q < q'$, and that $w^+(q, q') \leq v(q)$ for all q, q' satisfying $q > q'$. Less formally: relative to the demand curve of an inexperienced consumer, the demand curve of an experienced consumer whose consumption in the previous period was q' is shifted upwards to the left of q' and shifted downwards to the right of q'.

These assumptions are intended to represent the effects of status quo bias on consumer demand. The underlying idea is this: experienced consumers treat the consumption levels of the previous period as their status quo positions or reference points; preferences are conditional on reference points; with respect to changes in the neighbourhood of a reference point, substitution effects are very weak (i.e. indifference curves are highly convex). Thus, experienced consumers are reluctant to change previous levels of consumption, even if prices change. For entrants, in contrast, there is no scope for status quo bias, and so substitution effects are stronger. Notice that, in this model, reference points are adjusted every period. Thus, the length of a 'period' represents the time that it takes for a consumer to adjust his reference point in response to changes in consumption.

The WTA function describes the experienced consumer's current valuation of marginal units of G when he is consuming *less* than in the previous period; that it lies above the entry valuation function represents aversion to reductions in his consumption of G below the level of the previous period. The WTP function describes the experienced consumer's current valuation of marginal units of G when he is consuming *more* than in the previous period; that it lies below the entry valuation function represents aversion to reductions in his consumption of *other* goods below the levels of the previous period. If the graphs of the WTA and WTP functions are combined into a single graph (if necessary, linked by a vertical segment at the status quo quantity q'), the resulting curve, read from the vertical axis to the horizontal, is the demand curve for an experienced consumer. Status quo bias makes this curve highly inelastic at the status quo level of consumption (i.e. at the transition between the WTA and WTP functions).

Now suppose that supply of G is fixed: in each period, the quantity supplied is s, irrespective of the price. And suppose that, in each period, the price is at its market-clearing level. Let $p^* = v(s/n)$, i.e. p^* is the price at which each entrant's demand is a $1/n$ share of the total quantity supplied. It is easy to see that the model is in long-run equilibrium if in every period the market-clearing price is p^* and if, at that price, each consumer's demand is s/n. That is to say: if in any period t, the price is p^* and everyone consumes s/n, then in period $t + 1$ the market-clearing price is again p^*, and each consumer (whether an experienced consumer or an entrant) again demands s/n.

To explore the dynamics of the model, we have to 'seed' it by setting an arbitrary starting point. Let this be period 0. For the purposes of the investigation, I treat the consumption in period -1 of the experienced consumers of period 0 as exogenous. In particular, I do not require that the average consumption of these individuals in period -1 was s/n, as it would necessarily have been if the total supply in period -1 had been s and if the market had cleared in that period. Thus p_0, the market-clearing price in period 0, need not be at the long-run equilibrium level p^*.

Consider the case in which $p_0 > p^*$, and consider whether in period 1 the market would clear at $p_1 = p_0$. Notice that average consumption per person in period 0 must be s/n, and recall that the experienced consumers of period 1 are a random sample of the consumers of period 0. First suppose that $p_1 = p_0$. Then each experienced consumer in period 1 demands exactly the same quantity as he did in period 0. Thus, the average demand of experienced consumers in period 1 is s/n. But since $p_0 > p^*$, each entrant demands less than s/n. Thus, there is excess supply in the market: the market-clearing price must be less than p_0. Now consider whether in period 1 the market would clear at $p_1 = p^*$. Clearly, the average demand of experienced consumers will be at least as great if $p_1 = p^*$ as if $p_1 = p_0$. (Because the experienced demand curve may have a vertical segment, it is possible that the average demand is equal in the two cases.) So if $p_1 = p^*$, the average demand of experienced consumers is at least s/n, while the demand of each entrant is exactly s/n. Thus, either the market clears or there is excess demand. Combining the conclusions of this paragraph: $p_0 > p^*$ implies $p_0 > p_1 \geq p^*$. A symmetrical argument shows that $p_0 < p^*$ implies $p_0 < p_1 \leq p^*$. That is: if in some period the market-clearing price deviates from the long-run equilibrium price p^*, then in the following period the market-clearing price will move in the direction of p^*. Or in other words, long-run equilibrium is unique and stable. In real time, the speed of adjustment to long-run equilibrium depends on the length of a 'period', that is, on the rate at which consumers adjust their reference points. It also depends on the rate of turnover of consumers: the higher the value of r, the greater is the speed of adjustment.

This conclusion can be interpreted as follows. At any given time, experienced consumers value marginal reductions in consumption in terms of the WTA function while they value marginal increases in terms of the WTP function. The result is that marginal reductions have a higher absolute valuation than do marginal increases. In the long run, however, the market price is independent of the WTA and WTP functions; it is determined entirely by the valuations of entrants. Thus, it is only in the short run that status quo bias impacts on the market value of resources. For example, consider a project which is to start in some period t, and which will consume some fixed amount Δq of the good G in each period t, $t + 1$, ... Suppose that in period $t - 1$ the market was in long-run equilibrium, the price being $p^* = v(s/n)$. Let $p^{**} = v([s - \Delta q]/n)$, i.e. p^{**} is the long-run equilibrium price after taking account of the demand shift caused by the project. In period t, the market-clearing price will be greater than p** as a result of status quo bias. That is, the cost of the resources used by the project, measured at market prices, reflects the WTA valuations of experienced consumers. But as consumers adjust to the new conditions, the price will fall to p**; once this adjustment is complete, the cost of resources reflects the valuations of entrants.

Notes

1 A preliminary version of this paper was presented at the workshop *Projects, Programmes, Policies: Evaluation Needs and Capabilities*, held in Brussels in November 2000. It has been much improved as a result of suggestions and criticisms made at that workshop. I am also grateful to the Leverhulme Trust for supporting my research on the foundations of the theory of rational choice, which provides the background for the more specific work presented in this paper.
2 Further argument and evidence in support of the claims made in this and the preceding paragraph can be found in Sugden (1999a, 1999b).
3 These arguments are developed in more detail by Broome (1991).
4 This argument has been presented most forcefully by Sen (1992); see also Griffin (1986).
5 This and the following sections draw on my contribution to Chapter 1 of Bateman *et al* (2002).
6 Some commentators are convinced that CBA depends on such comparisons. For example, Broome (1999, p. 186) asserts that cost-benefit analysts 'add up utilities across people', adding 'This is not generally how they themselves describe what they do, but it is so'. On my account, it is not so.
7 To keep things simple, I leave aside issues of time discounting.
8 Strictly, this is true only in the limit as the size of the price change tends to zero and as the number of people affected tends to infinity.
9 This account of the market as a surplus-creating spontaneous order is expanded in Sugden (1998).
10 Serge-Christophe Kolm (1996) argues that utilitarianism is an Anglo-Saxon peculiarity. He has convinced me that Jules Dupuit, the nineteenth-century French economist who pioneered CBA, was not utilitarian in the Benthamite sense. It seems that Dupuit saw (what we now call) CBA as the measurement of economic surplus, and not as applied utilitarianism.
11 I say 'substantially' because some aspects of economic development and of social inclusion are external effects of transport projects, rather than indirect consequences of improvements in transport services.
12 I say more in defence of this suggestion in Sugden (1999a). I recognise that environmental public goods may have existence value as well as use value. I am not arguing that existence value can be ignored, but only that use value should be measured by means of a private-good payment mechanism.
13 Broome (1999, Chapter 12) distinguishes between 'structured' and 'unstructured' CBA, and argues for the former. Although Broome's account of the conceptual foundations of CBA is diametrically opposed to mine (he sees CBA as an investigation into goodness), his concept of structured CBA has the same *ex post* character as is required by my pay-as-you-go principle. He argues that a consistent application of the structured approach requires that CBA evaluates accident risks in transport projects from an *ex post* perspective. I am reluctant to accept this conclusion, for the reasons I explain in the main text.

14 A possible counter-argument is that, where the legal principle of strict liability is operative, private firms ultimately bear the *ex post* costs of risks that in the first instance are borne by their customers and employees. The uncertain prospects of current litigation by smokers to gain compensation from tobacco companies for health damage illustrate the ambiguities involved in the treatment of risk in the private sector.

15 In Sugden (1999a) I suggest that some *embedding effects* (i.e. cases in which individuals' stated valuations of public goods are unresponsive to the quantity of the good that is being valued) may be explained in this way.

16 One possibility, suggested by Bateman et al (1997), is to use survey designs which elicit *equivalent gain* (EG) valuations rather than WTP or WTA. Considering a benefit that takes some individual away from her status quo position, the equivalent gain measure of that benefit is the largest amount of money that the individual would accept *in place of* the benefit. Bateman *et al* find that EG valuations lie between WTP and WTA valuations.

References

Bateman, I., Carson, R.T., Day, B., Hanemann, M., Hanley, N., Hett, T., Jones-Lee, M., Loomes, G., Mourato, S., Özdemiroglu, E., Pearce, D., Sugden, R. and Swanson, J. (2002). *Economic Valuation with Stated Preference Techniques: A Manual*. Edward Elgar, Cheltenham.

Bateman, I., Munro, A., Rhodes, B., Starmer, C. and Sugden, R. (1997), 'A Test of the Theory of Reference-dependent Preferences', *Quarterly Journal of Economics*, Vol. 112, pp.479–505.

Broome, J. (1991), *Weighing Goods: Equality, Uncertainty and Time*, Basil Blackwell, Oxford.

Broome, J. (1999), *Ethics out of Economics*, Cambridge University Press, Cambridge.

Griffin, J. (1986), *Well-being: its Meaning, Measurement and Moral Importance*, Clarendon Press, Oxford.

Kahenman, D., Jack, L., Knetsch and Thaler, R. H. (1990), 'Experimental Tests of the Endowment Effect and the Coase Theorem', *Journal of Political Economy*, Vol. 98, pp.1325–1348.

Kolm, S.-C. (1996), *Modern Theories of Justice*, MIT Press, Cambridge, Mass.

Little, I.M.D. and Mirrlees, J. (1974), *Project Appraisal and Planning for Developing Countries*, Heinemann, London.

Pearce, D. *et al* (2001), *Guidance on Using Stated Preference Techniques for Monetary Valuation of Non-market Effects*, Manual prepared for UK Department of the Environment, Transport and the Regions.

Sen, A. (1992), *Inequality Re-examined*, Harvard University Press, Cambridge, Mass.

Sugden, R. (1998), 'Spontaneous Order', in P. Newman (ed.), *The New Palgrave Dictionary of Economics and the Law*, Macmillan, Basingstoke.

Sugden, R. (1999a), 'Public Goods and Contingent Valuation', in I. Bateman and K. Willis (eds), *Valuing Environmental Preferences: Theory and Practice of the Contingent Valuation Method in the US, EC and Developing Countries*, Oxford University Press, Oxford.

Sugden, R. (1999b), 'Alternatives to the Neo-classical Theory of Choice', in I. Bateman and K. Willis (eds), *Valuing Environmental Preferences: Theory and Practice of the Contingent Valuation Method in the US, EC and Developing Countries*, Oxford University Press, Oxford.

PART III
EVALUATION IN THE POLICY PROCESS

Chapter 10

Impact Assessment of Strategic Road Management and Development Plan of Finnish Road Administration

Eeva Linkama, Mervi Karhula, Seppo Lampinen and Anna Saarlo

Transport Policy Planning Framework in Finland

The Finnish Road Administration operates under the supervision of the Ministry of Transport and Communications. In 2000 the Ministry drew up guidelines for long-term planning of the whole transport sector, setting the target for the year 2025. The guidelines present a vision of the transport system, taking into account various policies and plans, adopted during the recent years by the Government. According to the guidelines of the Ministry, the objective of the transport policy is *intelligent and sustainable traffic and transport, which takes into account economic, ecological, social and cultural aspects.*

The guidelines set objectives – or targets – for the following five target areas:

- service level and costs of the transport system;
- health and safety;
- social sustainability;
- regional and urban development;
- negative impacts on the natural environment.

The guidelines are a policy paper, without immediate connection to annual finance or the strategic planning of *Finnra.* The only quantitative targets relate to traffic safety and air pollution. The rest of the targets are qualitative, mainly indicating in general terms the direction of the desired development or the desired state of affairs. The implications of the nature of the target-setting for road management and development planning are discussed in section 5 and 6 below, and the targets set by the Ministry are presented in detail in Annex 1.

The Ministry of Transport and Communications has also recently adopted a policy framework for environmental issues in the transport sector, setting general targets for *Finnra* and other national transport authorities. The environmental targets are more precise than the targets set for environmental issues in the general policy framework. However, the environmental policy framework does not include any quantitative targets.

So far no national multi-modal transport planning has taken place in Finland. For the first time the new strategy of the Ministry implies that in the future the Ministry will simultaneously evaluate all the major transport projects, covering roads, railroad, airports, harbours and terminals.

Transport and Roads in Finland

Finland is a sparsely inhabited, large country with long distances. The production and export infrastructure – such as forestry logistics – require effective transport from all corners of the country. Therefore goods transport per capita is relatively high. On the other hand, Finland has managed to keep the proportion of rail transport in goods transport two or three times that of the rest of EU. The share of using waterways is nevertheless lower, mainly due to the winter closure.

Finland has 2.4 million automobiles, including 2.1 million passenger cars. The population is in total about 5 million. The land area is roughly the same as that of Germany, making Finland one of the larger and most sparsely populated countries in Europe.

The difficulty of providing effective public passenger transport in large, sparsely populated rural areas makes the transport system structurally energy intensive compared with urban transport. This also applies to the goods transport and road maintenance. In total, 93 per cent of passenger traffic and 67 per cent of goods transport takes place on the roadways.

Roadways Network

Finnra is responsible for managing and developing the road network in rural areas as well as in towns and cities where major thoroughfares and arterial roads belong to *Finnra*. Municipalities bear responsibility for the local network in built-up areas, but also share costs with *Finnra* in developing the public road network in their jurisdiction. In addition to public roads and the municipal street network, there is an extensive private road network serving the needs of the forest industry, recreation or rural areas with a small permanent population.

Public roads	78 000 km
Streets	26 000 km
Private roads	280 000 km

Due to the low traffic volumes in most of the country, the total length of four- or six-lane motorways is only about 500 kilometres. About a third (in length) of the public roads are gravel roads. They will remain such in the foreseeable future.

The total length of the trunk roads (main roads class I and II) is over 13.000 km. More than 95 per cent are two-lane roads, which is clearly indicative of the nature of the network: long distances and small traffic volumes.

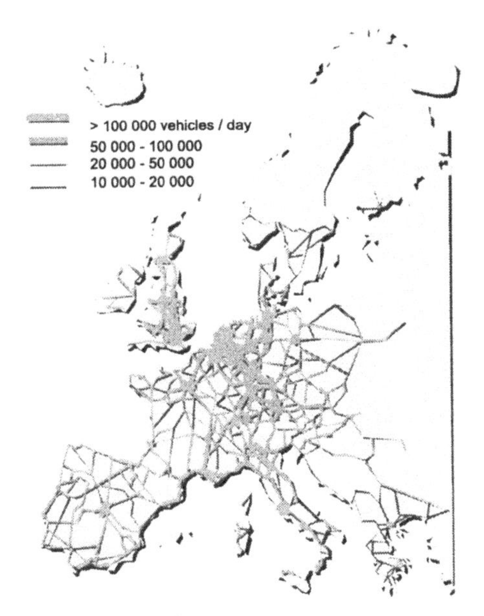

Figure 10.1 Traffic Volumes in Europe

Changes in the Finnish Society

Some major changes are taking place in the Finnish society, setting new requirements for road management and development. A growing share of the population, especially the young and well-educated, is moving from rural areas to towns and cities. This has made the Helsinki region the fastest growing metropolitan area in Europe at the present. The new information technologies are rapidly influencing the society, as new jobs mainly concentrate in major cities. The population is growing older, as elsewhere in Europe. All have major implications on traffic flows: these phenomenal growth concentrates around cities and on inter-city highways, rural roads become ever more quiet.

The fact that almost a third of the traffic volume on public roads is created in urban areas means that there will be growing capacity problems, albeit on only very limited sections of the road network, while the problems and deficiencies of the rest of the network relate mostly to safety, quality of pavement, winter maintenance etc. Taking into account the diminishing funding for road management and development, the key question becomes whether and to what extent to continue investing in the maintenance of the low-volume, secondary rural network, at the expense of major highway development.

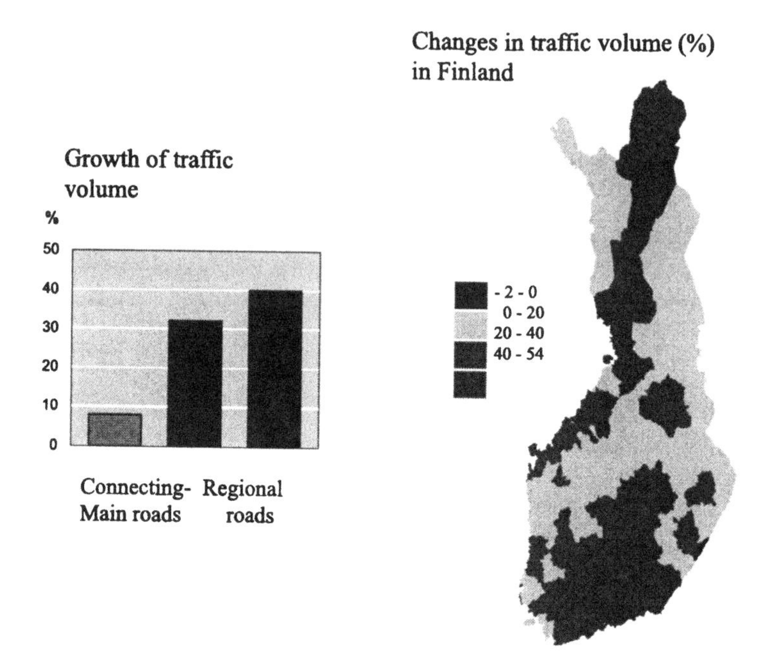

Figure 10.2 Traffic Forecast; Growth From 2000 to 2030

Road Management and Development Planning Framework

Funding

The funds for road management and development, as well as for railroad management and development, are allocated annually by the Parliament. There is no immediate connection between annual financial planning and targets for *Finnra*, as the annual targets relate only to selected operations of road management and development. The Road Administration distributes the funds to the nine road districts, which in turn allocate the funds for maintenance, upkeep and development of the road network in their region, based on the national and regional long-term plans and medium-term programmes. In excess of the annual funding for road management and development, the funding for major highway development projects is decided on a case-by-case basis by the Parliament. In order to support the Ministry of Transport and Communications in transport network planning, *Finnra* is preparing a national plan for trunk road development.

Planning and Programming in Finnra

Finnra prepares every four or five years a strategic road management and development plan for the following 10 to 15 years. Every year *Finnra* prepares an action plan (or programme) for the following five years, setting more detailed guidelines for road management and development and also defining major road projects to be commenced in the near future. Detailed programming takes place in annual budget plans and in regional planning carried out by the Road Districts.

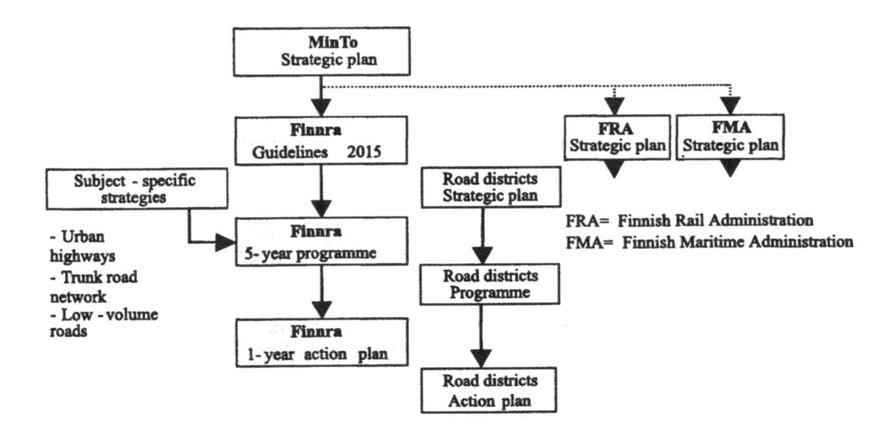

Figure 10.3 Road Transport Planning Framework in Finland

Based on the Guidelines for Road Management and Development, *Finnra* prepares three subject-oriented strategic plans: trunk road development strategy, urban highway development strategy and management of the secondary road network with very small traffic volumes.

Regional Strategic Planning

The national strategic plan is a guideline for road districts in their strategic planning. As the guidelines set the targets rather than a strictly defined framework, the road districts are in practice able to take into account regional needs and interests. The road districts are also in charge of defining more precisely the targets for traffic safety and environment. The strategic planning process, including participation, is strong at the regional level. The planning process of road management and development is, as described above, basically strictly hierarchical, but simultaneously flexible enough to provide space for regional needs and interests.

Strategic Environmental Assessment (SEA)

Aims and Principles of Strategic Environmental Assessment

Strategic environmental assessment (SEA) can be defined as the formalised, systematic and comprehensive process of evaluating the environmental impacts of a policy, plan or programme and its alternatives, including the preparation of a written report on the findings of that evaluation, and using the findings in publicly accountable decision-making. (Partidário, 1999). According to another definition, SEA aims to provide the competent authority with a tool which enables it to be fully aware of the environmental/sustainability issues associated with a particular policy, programme or plan *while the latter is allowing that it is being formulated* or, at the very least, evaluated before it is implemented. SEA aims to expand the competent authority's focus by providing a mechanism by which its goals and objectives will include cross-cutting environmental and sustainability perspectives (Partidário, 1999).

SEA, or more accurately *impact assessment of strategic plans*, has been widely discussed by environmental impact assessment experts throughout the world. A widely accepted conclusion is that the methods and practices of SEA are only evolving, so far without established procedures. One of the key findings is that in SEA *the process is more important than the final product on paper.* In other words, impact assessment of strategic plans achieves results only when impact assessment is an integral part of the process. The idea is to 'guide' through the preparation process of the plan with help of the results of impact assessment. The rationale behind this is simply that the number of alternatives in strategic planning is theoretically unlimited. Also, cumulative and indirect effects of strategic plans may be more important than separate findings of impacts on single issues.

One rather extreme view considers impact assessment as *a politicised process that improves decision-making* (Barrow, 1997). Politicised or not, when applied wisely, the results in improving decision-making can make SEA a valuable tool. Barrow has recognised several difficulties of applying impact assessment above project level (to policies, plans or programmes):

- plans often relate to a number of possible developments in various places, which makes matters much more complex than when there is a single project focus;
- in view of the above, it is difficult to be precise in predictions, which makes impact assessment much more of a challenge;
- there is less hindsight experience in the impact assessment of policies, programmes and plan than there is at the project level;
- scoping at levels above project impact assessment is difficult; what should be focused upon?

According to Partidário (1999) SEA can only promote sustainability if there exists:

- a policy framework linking with other policy tools and institutional contexts;
- credible and feasible alternatives that allow evaluation based on comparable rather than absolute values;
- recognition of the uncertainty that characterises any policy and planning development context;
- simple but pragmatic indicators that can assist monitoring of the assessment process;

- good communication mechanisms to ensure that all partners in the SEA process are adequately involved and their perspectives considered.

How well these conditions were met in *Finnra*'s exercise is discussed below.

SEA in Finland

The Act of Environmental Impact Assessment Procedure took effect in 1994. The Act stipulates that when an authority prepares a plan, programme or policy whose implementation is likely to have a significant impact on the environment, its impact must be investigated and assessed *to the necessary extent*. The Government (Council of State) issued in 1998 general guidelines for assessing impact of policies programmes and plans. The aim of the guidelines is to support better compliance with the assessment obligations and to promote the consistency of the assessment practices.

The concept of 'environment' is defined in Finland in broad terms: beside living nature and natural resources, the concept also includes human health and living conditions.

Environmental impact refers to the various direct and indirect consequences of human actions, both in Finland and beyond its frontiers. This is taken to include actions which affect

- human health, living conditions and public amenity;
- the soil, water, air, climate, flora, fauna, interactions between them and biological diversity;
- community structure, buildings, the landscape, townscape and cultural heritage;
- the utilisation of natural resources;
- the interaction between the above mentioned.

So far there are no established methods or practices in assessing impacts of plans and programmes in Finland. The new EU directive on SEA and the upcoming national legislation will define the practices in the near future. However, strategic environmental assessment has been for the first time systematically applied in *Finnra* in the planning process of the Guidelines for Road Management Development 2015 and in strategic planning of one of the road districts, Savo-Karjala.

Guidelines for Road Management and Development 2015

Nature and Purpose of the Plan

The development of the society places many expectations on road management. These national and regional needs constitute a starting point for road management and development planning. Customer orientation and societal responsibility are the most important values of road management and development. The quality of products and services, their availability and timing are based on awareness of the goals, needs and expectations of the customers.

The plan is based on realistic expectations on the future finance of road management and development. It is expected to reach two goals:

- the Road Administration wishes to set out to what extent, within a given level of finance, it is able to meet the goals and expectations set by the society;
- internally the plan is a guideline for road districts' strategic plans and the coming five-year programmes of *Finnra*.

Previously the strategic planning in *Finnra* placed the emphasis on network development. Simultaneously, the planned level of funding was based on unrealistic expectations. Yet, the appropriations for road management and development strongly decreased since the early 1990s from the top level of about 1000 million Euro a year down to about 700 million Euro at present. Therefore, the plans were not credible, consequently not reaching the kind of steering role for medium-term programming they were expected to reach. In fact, the ultimate purpose of a strategic plan is to steer more detailed planning or programming and ultimately the actual operations of the organisation.

Given this unsatisfactory situation, *Finnra* set the following goals for the new plan:

- the plan is a tool in steering more detailed planning and programming in *Finnra*, including the plans and programme of the road districts;
- therefore, the plan must be credible, based on realistic expectations on future finance. (The plan was based on the present level of finance, about 700 million Euro annually);
- the plan takes into account various regional needs and expectations in the national framework;
- participation of national and regional stakeholders is an integral part of the preparation of the plan;
- the plan takes into account the interests of various road-users on different parts of the road network.

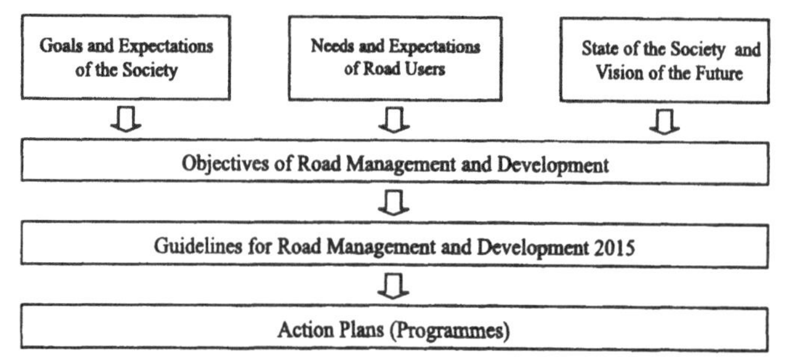

Figure 10.4 Road Management and Development Planning in Finnra

An overriding political guideline in preparing the plan is the requirement to keep the whole network of public roads open and in a satisfactory condition all year round, 24 hours a day.

Objectives for Road Management and Development

The targets set by the Ministry of Transport and Communications include no quantitative elements, making it difficult to interpret them in the strategic planning process. In relation to the implementation of the targets, another problem is caused by the fact that the Ministry has not given priority to any of the targets. Therefore, *Finnra* found it necessary to conduct unofficial consultations with the Ministry with regard to the interpretation and prioritisation of the targets. *Finnra* subsequently came to the conclusion that it should be able to meet all the targets in a satisfactory manner. It was left to *Finnra* to define what is meant by 'satisfactory'.

As the combined result of the targets of the Ministry and needs and expectations set by the society, the objectives are more extensive than before, reflecting the pluralistic development of society. In part, the objectives are contradictory as well. This will entail a greater challenge for road management planning and highlight the importance of a broader expertise and co-operation.

The objectives of road management and development relate to six main target areas:

- economic efficiency in broad terms, including road users, the government and the society as a whole;
- operational requirements of the business society;
- regional equality;
- social equality;
- safety;
- environment.

Each target area includes three to five separate objectives (see Annex 2).

The targets set by the Ministry could not be applied as such in the strategic planning of *Finnra,* due to their general nature. Therefore, the objectives were re-formulated by the road management experts of *Finnra.* The process of formulation of the objectives is discussed in further detail below. Based on the expectations expressed by the various stakeholder groups, three objectives were identified as more important than the others: traffic safety, accessibility of the whole network and satisfactory upkeep of the network. It is also worth noticing that more emphasis than before was placed on environmental issues as well as on regional equity.

Planning Process and Impact Assessment of the Guidelines

In the road management and development plan the conditions for promoting sustainable development, as described in the previous chapters, were met in the following manner:

- the plan is based on the policies of the Ministry of Transport and Communications, other relevant policies as well as on the policies and programmes of the Council of State;
- the plan is based on impact assessment of three alternatives and on comparing the alternatives;
- impact assessment was characterised by qualitative analysis, instead of quantitative analysis that has traditionally been applied in transport planning. Qualitative analysis tends to be more sensitive to uncertainty than quantitative analysis;
- the key goal in defining indicators was to ensure the availability of relevant information;
- participation of stakeholders was included in the planning process right from the beginning.

Experience (Thérivel and Brown, 1999) suggests that strategic impact assessment (SEA) is often most effective when carried out by a multidisciplinary team from the competent authority, with input from a SEA consultant. *Finnra* selected this approach while preparing the plan. While the experts of *Finnra* have deep knowledge of the issues and impacts of road management, the consultant was able to bring into the process a relevant methodological approach and to manage the assessment process. The phases of the assessment process, integrated into the preparation of the plan, are presented in a chart (Annex 3). It is noteworthy that the plan discusses the upkeep and maintenance of the road network with even more emphasis than new road projects. This is mainly due to the fact that accessibility and upkeep of the whole network are prioritised. The shift in emphasis is even more important now that the appropriations for road management and development do not reach the level of the late eighties and early nineties. The lower level of funding leaves less room for investment, simultaneously making it ever more important to keep the existing network in a satisfactory condition. By using the present level of finance as the basis for the plan, *Finnra* is able to present not only what can be accomplished but also what cannot be accomplished in road management and development.

Goal Model

In assessing impacts of the Road Management and Development Plan *Finnra* applied the goal model (also called the goal-attainment model). The idea of this model is to assess to what extent the goals (or objectives) are attained and what other impacts (side effects, usually negative) the plan has. The goal model is based on the assumption that general targets set by the society can be formulated in such a way that the path from the present to the future can be presented. The model attempts to reach an optimal solution or a solution the impacts of which are acceptable.

The major difficulty in using the goal model is the general nature or vagueness of the goals (or targets). This is also the case in strategic road management planning, as described earlier. If the goals can be interpreted in more than one way, it is impossible to formulate the future target situation to be accomplished by implementing the plan. One way to interpret the goals is to use indicators to measure the level of accomplishment of the goals. The vagueness of the goals can promote public discussion on goals. Discussion is a positive factor both from the perspective of assessing the plan as well as that of assessing the goals.

The tradition in mainstream transport planning has been almost solely problem-oriented, regardless of the way of defining problems. Problem-orientation has been the approach in previous strategic plans of *Finnra* as well. In spite of using the goals as the basis for the plan, the traditional approach has to a certain extent had a role in the new plan. Problem recognition always plays a role in transport planning, but to a lesser degree than in the past.

Methods of Assessment

Qualitative Analysis

The objectives of the plan mostly define the *direction* of change instead of defining exact targets. The objectives include expressions such as *diminish* negative effects on ground water, *improve* conditions of public transport users, bicyclists and pedestrians in urban areas or *develop* international transport corridors.

Taking into account the nature and formulation of the objectives, there is very little space, or need for that matter, for quantitative analysis. The most important thing is to bring about a common interpretation of the goals and a common judgement of goal attainment. Therefore, goal attainment can be judged by common expert view. The whole team of road management and development experts of *Finnra* participated as a team in assessing the impacts of the plan and consequently goal attainment. The team represents all aspects of road management and development, making it possible to use in-depth expertise in formulating a reliable vision of the impacts of the plan and of the future situation as a result of the implementation of the plan.

Judging goal attainment requires individual expert view on the subject matter. Consequently, the present situation has to be analysed, and the relevant factors affecting attainment of the specific goal have to be defined. The expert view was solidly based on the R&D activities of *Finnra* as well as previous planning. Also, the results of impact assessment of major road projects were utilised in the process.

Objectives

The objectives of the road management and development plan are to a certain degree contradictory. As the set of objectives include e.g. environmental issues, it is very difficult, or even impossible, to simultaneously achieve the objectives regarding development of the main highway network and environmental objectives.

For this reason, it was found necessary to divide the objectives into three groups, based on their importance and the extent to which the plan is effective in attaining any specific objective. To give an example of the latter perspective, the plan is expected *to provide support for the development of the transport system as a whole so that the transport system would support the development of cohesive urban structure to diminish the need to use automobile and to improve possibilities to use public transport*. No matter how important the objective is, it was recognised that the plan is not very effective in attaining this objective.

As a conclusion, six goals were found to be of 'utmost importance', three goals 'very important' and the rest (twelve) 'important'. The results of assessment were classified into five groups based on what extent each alternative either supports or harms attainment of each specific objective (strong support – weak support – no noticeable effect – little harm – serious harm). The classification was a common (qualitative) interpretation of the road management and development experts on how well the objectives are attained. Also the uncertainties and risks related to the impact assessment were evaluated.

C/B-Analysis is Missing?

No cost-benefit analysis as such was included in the assessment process. This is evident because of the nature of the plan, not being an investment programme. It was concluded that C/B-analysis is not a relevant method in strategic impact assessment because of its structural weaknesses.

C/B-analysis is a quantitative method, which can and should be applied in more detailed planning and programming, where the objectives are mostly quantitative and well-defined. In the planning framework of *Finnra* this means that C/B-analysis can be used – and is actually used – in three-to-five-year programming and in activity-based strategic planning, e.g. for trunk road network development or network upkeep planning.

It was also concluded that he 'technical approach' of C/B-analysis would have been counterproductive in the planning process, as the emphasis was laid on strategic questions and policy issues, purposefully diverting discussion from quantitative issues and highway development projects. Taking into account the vague nature of targets of the Ministry of Transport and Communications, it would have been very difficult, if not impossible, to formulate a politically satisfactory set of quantitative goals and indicators required for a C/B-analysis.

However, it must be noted that all the major (in this plan undefined) highway projects which will be realised during the planning period must pass the requirement of a C/B-ratio of 1.5 (equalling net ratio of 0.5). In relation to available finance, there is in fact an overflow of possible highway projects which meet the requirement. *Finnra* also applies the Pavement Management System (PMS) in evaluating necessary actions in pavement upkeep. Based on C/B-analysis, the method defines the optimum condition of pavement and the timing of necessary upkeep (repavement or other actions).

In a very sparsely populated country like Finland with low traffic volumes, it is not possible to base the level of maintenance on optimising costs and benefits in narrow economic terms. About 50.000 km of the total network of 78.000 km are secondary roads, serving rural residents and businesses. Accessibility all year round and adequate traffic safety are considered basic rights, the value of which is very hard, if not impossible, to estimate. As a matter of fact, maintaining accessibility on public roads is a unanimous political choice.

Alternative Plans

There were two basic rules in formulating alternative plans: first, they must be distinctive enough to make comparison relevant, and second, they must be realistic enough so that any one could be the basis for the guidelines. In the early phases of the preparation of the plan three separate approaches in formulating the alternatives were discussed. One of the discarded approaches was based on the regional approach (emphasis on growing cities – other cities and towns – rural areas), the other on the classification of the road network (emphasis on urban areas – trunk roads – other parts of the network).

The selected approach is based on the distinction between various road user groups. In analysing different approaches it was found that in the road-user approach it is possible to take into account regional aspects as well as various parts of the road network. The approach can also be seen as a manifestation of customer orientation in road management and development. The first alternative put emphasis on serving car traffic on main highways, the second alternative on significantly improving the conditions of pedestrians, bicyclists, and public transport users, and the third alternative on guaranteeing adequate level of service to goods transport and other transport needs of the business community across the country.

The alternatives were characterised by the allocation of funds on various 'products' of road management and development (maintenance, upkeep, construction, telematics and services) as well as on various parts of the road network. Altogether 12 variables were used in formulating the alternatives:

- maintenance of the network;
- upkeep of the network;
- major trunk road network development;
- other trunk road network development;
- development of urban highways;
- development of public roads in small towns and villages;
- development of other parts of the network;
- traffic safety improvement;
- actions to diminish negative environmental impacts;
- improvement of conditions of pedestrians and bicyclists;
- replacement of ferries;
- traffic management and telematics.

The rationale of using alternatives in policy formulation is related to the discussion above on the 'vagueness' of goals. Alternative policies form a rudimentary framework for assessing goal attainment, making it possible to evaluate to what extent the changes in funding (between various 'products' of road management) and changes in allocating funds on various parts of the road network actually affect the attainment of goals. The total amount of finance was kept equal in all three alternatives. In all three alternatives the share of maintenance is about 30 per cent.

The comparison of the alternatives was carried out by the team of road management and development experts of *Finnra*. The comparison was based on a disaggregate approach. In the early phase of assessment some tests were made to apply aggregate methods. Very soon it was found that it is unfeasible, if not completely impossible, to combine aggregate comparative methods, using money, points or other ways, with qualitative analysis. The objectives of 'utmost importance' (see earlier) were emphasised in comparing the alternatives. It was considered unacceptable that the final plan would have included policies or actions that would have seriously harmed attainment of any of the most important objectives.

Participation

The purpose of stakeholder participation in the planning process was on the one hand, to get a clear picture of stakeholder expectations and, on the other hand, to get feedback from stakeholders during the preparation of the plan. *Finnra* wanted to find out clients' and partners' goals and expectations in order to come up with well-balanced guidelines.

Who are the stakeholders of a national strategic plan? The easiest, if not very practical, answer is each and every citizen. Another, hardly more practical, way to define the stakeholders is partnership – those who have interest in participating. In practice, the stakeholders are national and regional interest groups and authorities.

As Tomlinson and Fry (2000) note, there are numerous practical issues in stakeholder involvement that are not found with project-level environmental impact assessment (EIA). The most important difference has to do with the role of the general public, which has a major role in EIA but a minor role in SEA. A major obstacle in organising the participation of the general public is that plans and programmes are likely to be too remote to interest people other than members of the interest groups.

In the preparation of the road management and development plan of *Finnra* regional stakeholders included established partners of road districts, such as regional associations of municipalities, environmental authorities, chambers of commerce, regional representatives of national government organisations, associations of trucking companies, nature preservation associations, major industrial enterprises etc. The road districts made independent decisions on organisations to be invited to participate in the process. On the national level stakeholders included various ministries, other government organisations in the transport sector, national associations of various interest groups etc. Participating organisations had the full freedom to nominate their own representatives in the process.

Methods of participation included introductory seminars, providing an opportunity to comment on the preparation of the plan, and written comments. National and regional seminars were meant to provide an opportunity for stakeholders to present their views in different phases of the planning process. No formal public hearings were organised.

In the early phases of the preparation of the plan much emphasis was placed on introducing the nature of the preparation process as well as covering stakeholder expectations. The information was especially utilised in formulating objectives and alternative plans. During the later phases of the project the draft plan was introduced and discussed. Participating stakeholders were asked to provide written comments on the draft. Written comments varied from broad strategic issues to minor details of the formulation of the plan, all contributing to the finalisation of the plan.

Social Impact Assessment

Social and regional equality was included in the objectives of the plan. However, the concept of equity has been operationalised neither by the Ministry of Transport and Communications nor *Finnra*. So far, methods of social impact assessment have been very little studied and discussed in strategic planning. The difficulty lies in the nature of the plan, which does not include location-bound projects or programmes relating to any of the road management and development activities. Therefore, social impacts were assessed as a cumulative outcome of various impacts. The assessment was backed by the allocation of funds in different alternatives on various road management and development activities. The results of social impact assessment were described as impacts on various road user groups on different segments of the road network by using the classification presented in Figure 10.5.

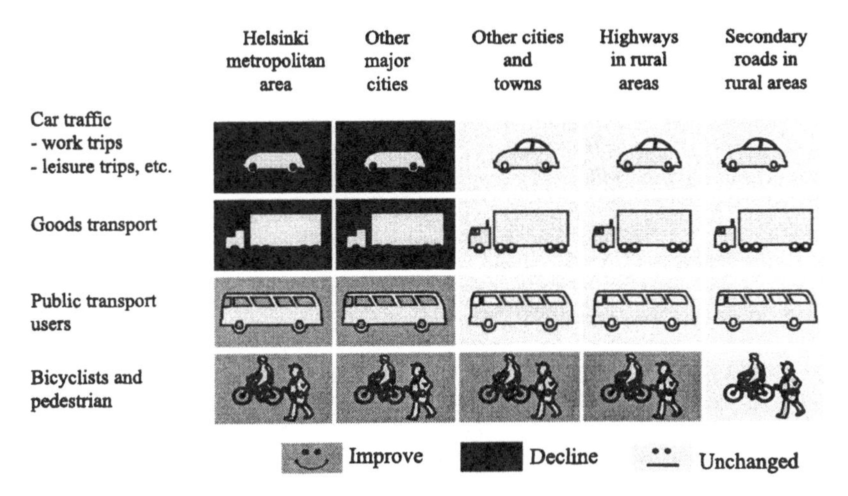

Figure 10.5 Framework for Assessing Impacts on Various Road Users

Finalisation of the Plan

The conclusion of the assessment and comparison of the three alternative plans was that no alternative was acceptable as such. Therefore, the final plan was a compromise, aiming at making necessary adjustments to eliminate major weaknesses without losing the best qualities of the three alternatives. This goal was achieved satisfactorily.

The final plan was put together in such a way that no serious harm to the attainment of any goal of 'utmost importance' was accepted. Technically the task of finalising the plan can be described as optimising the allocation of funds. Another way of looking at the task is to describe it as a process of finding the best possible compromise among conflicting goals and interest.

The impacts of the final plan were assessed by using similar methods as in assessing the three tentative plans. As Barrow has noted, it was soon concluded that it is very difficult to be precise in predictions. The plan relates to a number of possible developments in various places and various sections of the road network, without defining the exact nature and location or section.

In practice the final plan was created during the planning process along with the increase of information on the impacts of the alternatives and the feedback from the stakeholders. Following discussion by the Board of *Finnra*, the draft plan was sent to participating national and regional stakeholders for written comments. *Finnra* received comments from about 30 national and about 80 regional stakeholders.

The road districts were responsible for sending the plan for comments and for summarising the comments of the regional stakeholders. The policy of the road districts varied considerably in this respect: the biggest number of stakeholders who were asked to submit their comments was 40, the smallest number two. It may be concluded that the latter example did not accomplish the aims set for stakeholder participation.

The final decision on adopting the plan was made by the Board of *Finnra*. In the Finnish transport planning framework there is no other decision-making procedure, e.g. by the Ministry of Transport and Communications.

Documentation and Audit

The Guidelines for Road Management and Development 2015 are documented in three publications:

- summary (also in English);
- main report;
- background material, including the external audit report.

The background material is an extensive publication on the planning process, impact assessment and participation, aiming at making the process transparent to anyone interested in how the plan was prepared and formulated. In addition, other unpublished material was made available to the external audit group.

After finalising the plan, the planning process and impact assessment was audited externally. A contributing factor in organising an ex-post evaluation was the feedback from the stakeholders, suggesting that *Finnra* would also use independent experts in assessing the impacts of the plan. The audit was carried out by an independent group of experts, representing Tampere University of Technology (traffic and transport planning), the Finnish Environmental Centre (environmental impact assessment), the Research Centre for Health and Social Affairs (social impact assessment) and the Finnish Technical Research Centre (economy). The audit group was specifically asked to comment on the planning process, the impact assessment process and the results of the impact assessment.

The comments of the audit group were listed under the following headings:

- preparation process and impact assessment as a whole;
- formulation of alternative plans;
- goal model in impact assessment;
- credibility of impact assessment;
- (missing) role of political decision-making.

The group concluded, after commenting and criticising various parts as well as details of the process, that the impact assessment had been a learning process and, in that sense, much more than a single project. The group recommended that the road districts of *Finnra* adopt a similar process in order to support regional strategic planning.

Experiences

Client Expectations and Participation

Because of the in-built problems, strategic planning has been developed towards communicative planning, meaning wider and deeper participation during the planning process. While preparing the plan, *Finnra* aimed at reaching new levels in organising participation of various stakeholders. Based on the report of the external audit group, the goal of more extensive participation was reached satisfactorily. Admittedly the methods regarding participation need to be developed further. On the other hand, the feedback from the stakeholders was positive.

As international experience implies, it was found more useful to involve representatives of public interests rather than attempting general public consultation. The general public being the whole nation, it would have been very difficult to find any practical and meaningful ways to organise the process. The interests and needs of various road user categories in different parts of the country and on different parts of the road network played a bigger role than before. The plan is also a link to regional road management and development planning.

Objectives

During the impact assessment process, it was found that more emphasis should have been placed on defining the objectives of road management and development. Many of the objectives were too vaguely defined, leaving too much space for interpretation. The vagueness of the objectives also caused much discussion among the experts of *Finnra*, sometimes making it difficult to agree on the results of impact assessment.

It was found that there were too many objectives. As the impact assessment was based on the goal attainment model, no objective could be neglected, no matter how vague from the view point of strategic road management and development planning. Even though the objectives were classified into three groups, it was sometimes difficult to keep discussion – even among the experts – on key issues.

It was clearly recognised that setting priorities on objectives had very much to do with values and attitudes. It is in fact that whatever the assessment method, values and attitudes always play a role – smaller or bigger – in preparing a strategic plan. The question is whose values are emphasised. A partial answer to this problem is transparency of the process and real open-minded interchange with clients and partners.

Baseline Information

It was found that the baseline information was not adequate in all respects. Even though the level of information was very satisfactory for most purposes, more information is needed on travel patterns. On the public road network most of the traffic is created by short-distance leisure or other non-work travel, shifting emphasis from clearly-patterned long-distance travel, creating numerous small dispersed traffic flows. In order to support public transport, to reduce negative environmental impacts and to improve traffic safety, more research is needed.

Social Impact Assessment

Not surprisingly social impact assessment was found to be the most difficult part in impact assessment. Participatory methods are practical in impact assessment of strategic plans, but in this case they were not applied. This was partly due to the fact that the impact assessment was added to the process only at a later phase.

Impacts of Impact Assessment on the Plan

The results of impact assessment did influence the way the final plan was put together. However, it is evident that during the final stages of the preparation, and even more during the decision-making phase, various interests tend to become more apparent, impact assessment being only one factor affecting the final outcome. While the plan was finalised, political realities were taken into account, concerning e.g. regional equity and the extent of the public road network of the lowest traffic volume as well as the formulations of the politically delicate motorway network development.

Sensitivity of Impact Assessment

The sensitivity analysis concluded that the plan is not very sensitive to changes in traffic forecast. This is not surprising given the low traffic volumes and nearly non-existence of capacity problems. Instead, the approach is sensitive to the values and attitudes towards the objectives. It must be noted that the values and attitudes of *Finnra's* road management experts most likely have a certain impact on the results of impact assessment, as assessment was mainly based on qualitative methods.

Ex-post Evaluation

Finnra commissioned an ex-post evaluation of the planning process to an independent consultant. The results of the evaluation were summarised in a number of recommendations for development of strategic planning, covering participation, transparency of the planning process, utilisation of *Finnra*'s personnel resources, internal communication and further development of follow-up procedures as well as the structure of national of regional road management and development planning.

Further Steps in Strategic Planning

Planning and Programming

In the next phase the road districts will prepare their strategic road management and development plans. *Finnra* is preparing three more detailed strategic plans to complete the Guidelines for Road Management and Development:

- guidelines for trunk road development;
- guidelines for urban highway development;
- guidelines for management of the secondary road network with very small traffic volumes.

The trunk road development strategy will discuss the extent and nature of the highway development, on one hand defining nation-wide standards for highway improvement and, on the other hand, determining the road sections to be developed during the next 10–15 years.

The urban highway development strategy will redraw the policy of *Finnra* in major urban areas. The policy will take into account the new guidelines for road management and development particularly relating to public transport as well as pedestrian and bicycle traffic. There has been criticism towards earlier policies, based very much on *Finnra* using design standards unsuitable for urban areas. Guidelines for management of the secondary road network will set nation-wide standards for maintenance and upkeep of low-volume secondary road network.

During the next round of planning *Finnra* will prepare a five-year programme, based on the new guidelines. Correspondingly, the road districts will prepare their programmes based on national and regional strategic plans (guidelines). In this way the plans will be tiered to provide a coherent framework for further planning and design as well as for daily operations of upkeep and maintenance. The guidelines will be redrawn in about three years, either based on the existing plan or from scratch, depending on the future development generally, but very much also on possible changes in the financial framework of *Finnra*.

There will be no separate ex-post evaluation of the implementation the new guidelines. On the level of strategic long-term planning there are, in the period of five or ten years, no practical ways to distinguish the impacts of the plan from other factors affecting road management and development. A true test of the guidelines will be the success of tiering various plans on the national level with five-year and annual programmes as well as with regional guidelines and regional programmes.

Conclusions

De minimis non curat lex (The truth does not lie in the detail)

For the first time in Finnra's long-term strategic planning history the plan is resource-based and goal oriented, a genuine strategic plan. Earlier even long-term plans tended rather to be investment programmes. This time the discussion very much emphasised strategic issues. However, whether the successful completion of the planning process guarantees successful implementation of the plan is another matter. Road management and development is a deeply political issue, leaving any plan open until final decisions of the implementation of road construction projects have been made or maintenance contracts have been signed.

Regarding the planning process, the goal model was found a satisfactory method in assessing the impacts of the plan. The key issue in applying the goal model is the success in defining objectives in such a manner that impact assessment rests on a solid basis. If the objectives are too vague, there will be an ever-lasting dispute on their correct interpretation. If they are contradictory, as they often are, almost any conclusions can be based on them. If the objectives are unrealistic, no real plan can be based on them.

Qualitative analysis is not the traditional planning approach in the transport sector. Yet, there is no need to move towards quantitative analysis in strategic planning. Whenever quantitative analysis is found useful, it is important to make sure that the quality of baseline information is adequate for strategic planning. As an entirely quantitative approach, a cost-benefit-analysis would not have been satisfactory in the sense of directing the discussion on strategic issues of road development and development.

Despite all the difficulties mentioned in this article, *Finnra* is determined to further develop the application of the goal model in strategic planning. So far there seems to be no serious challenge to the goal model. A guide on impact assessment (or strategic environmental assessment) for future strategic planning processes, based on the experiences and lessons learnt, will be prepared for *Finnra*'s internal use.

References

Barrow, C.J. (1997), *Environmental and Social Impact Assessment; An Introduction*, Arnold, London.

Partidário, M.R. (1999), 'Strategic Environmental Assessment – Principles and Potential', in Petts, Judith (ed.), *Handbook of Environmental Impact Assessment, Vol. 1, Environmental Impact Assessment, Process, Methods and Potential*, Blackwell Science, Oxford.

Thérivel, R. and Brown, A.L. (1999), 'Methods of Strategic Environmental Assessment', in Petts, J. (ed.), *Handbook of Environmental Impact Assessment. Volume 1, Environmental Impact Assessment: Process, Methods and Potential*, Blackwell Science, Oxford.

Tomlinson, P. and Fry, Ch. (2000), 'SEA and its Relationship to Transportation Projects', Paper presented at TransTalk Workshop on 30 October, Brussels.

ANNEX 1

Targets for the Transport System Set by the Ministry of Transport and Communications

Target areas	Target
Service level and costs of the transport system	Movement of people and goods should be safe, moderately priced and of high quality. All regions should enjoy the same basic level of mobility. Both domestic and international passenger and freight services should be reliable and smooth. Transport information should be reliable, easy-to-use and up-to-date. The transport system should be developed and maintained in a cost-effective manner. The passenger and freight transport markets should be efficient and open to competition. The Finnish transport sector should be competitive on both domestic and international markets.
Health and safety	The transport system as a whole should support an improvement in people's health. Nobody should have to die or suffer serious injuries in traffic.
Social sustainability	The benefits and negative impacts of transport should be evenly and fairly distributed amongst different population groups. Special consideration should be given to the needs of vulnerable groups. Individual citizens should be able to participate in and influence the traffic planning process.
Regional and urban development	Regional development targets set at national level and the regions' own development strategies should be supported by the transport system. The targets concerning urban structure and cityscape should be supported by the transport system. The transport planning and land use planning processes should be compatible and consistent with each other. Traffic environments should be pleasant and safe.
	The cityscape and cultural and historic landscape shall not be altered unless there are strong reasons to do so.
Negative impacts on the natural environment	Both global and local negative impacts on the natural environment shall be minimised. Use of natural resources (such as energy, soil materials and land) shall be minimised.

ANNEX 2

Objectives of Road Management and Development

Objective areas	Objective
Socio-economic efficiency	Target and optimise road management actions in an effective and economical way. Ensure road network condition and daily trafficability in the entire country throughout the year. Provide the prerequisites for transport chains between different means of transport. Support an integrated and solid societal structure. Support and promote the goals of national and regional traffic systems and implement the relevant investments.
Operational requirements of business life	Provide feasible and safe main road connections and routes to harbours and trucking terminals. Ensure necessary transport on the lower-standard network throughout the year. Enhance the safe and smooth flow of traffic and reliability of transport by traffic management. Develop international transport corridors and traffic areas.
Regional equality	Take account of the national land use objectives and promote their implementation. Take account of the needs of the population and business life in different areas.
Social equality	Ensure smooth passenger traffic for all population groups.
	Improve conditions in public transport and bicycle and pedestrian traffic, especially in major urban areas.
Road safety	Improve traffic safety so that fatalities and injuries in road traffic will decrease. Prevent off-the-road accidents and head-on collisions and alleviate their consequences. Enhance safety among pedestrians and bicyclists and reduce accidents in slippery conditions.
Environment	Promote the implementation of national climate strategy in order to reduce carbon dioxide emissions in road traffic. Reduce groundwater pollution and noise resulting from road maintenance and traffic. Adapt road traffic solutions to the urban and suburban structures. Promote the preservation of important nature reserves and cultural heritage areas. Promote biodiversity and economical use of natural resources.

ANNEX 3

Impact Assessment of Guidelines for Road Management and Development 2015

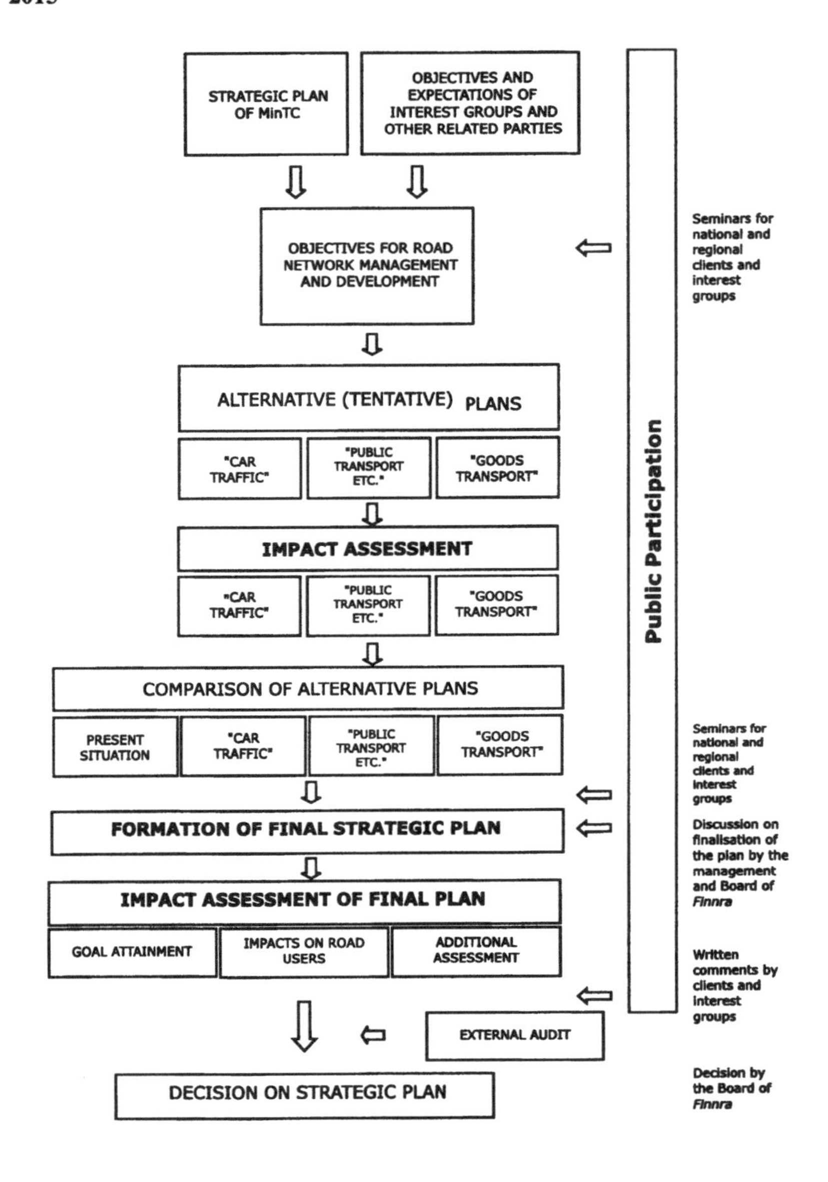

Chapter 11

Major Infrastructure Transport Projects Decision-Making Process: Interactions between Outputs and Outcomes as a Contemporary Public Action Issue

Marianne Ollivier-Trigalo

Decision-making in the field of major infrastructure transport projects is complex in our contemporary occidental societies. This statement, shared by analysts, decision-makers and other experts, is based on the recurrent conflicts that those processes inevitably produce. Planners and decision-makers still dream about an enchanted world, the one they think that they have lost: that in which they, together with some lobbies, were successful in building every conceivable piece of infrastructure for the welfare of the community. But times have changed (if they were ever thus). The social sciences can help approaching the new constellations of actors and their interaction in the decision-making process.

Two new actors have entered the decision process more actively during the last twenty years. One category is that of individuals who set up associations; the other is green parties. Today, the complexity of decision-making is characterised by the expression of a new rationale (versus the technico-economic rationale of planners) and by the appeal for accountability for the consequences of public actions, particularly with regard to environmental protection.

Concepts and scientific tools coming from public policy analysis and the sociology of public action are here relevant. The major transport infrastructure projects are considered as part of public policy and approached as a social system (Thoenig, 1985). With such a viewpoint, the decision-making process is shaped by a system of interacting actors who employ certain practices or activities to reach their objectives (Crozier and Friedberg, 1977; Friedberg, 1993). These practices and activities depend on the structure of the power relationships, notably through the institutional framework (Padioleau, 1982). The players involved in the process configure the relevant issues and, through their interaction, the rules of the decision-making process.

The analysis of conflicts that occur during any major infrastructure project have highlighted how public action and policy were shaped by multiple actors, contradictory interests and multiple finalities (Ollivier-Trigalo, 2001). The strategies of the different actors involved in such processes are also determined by their own specific social worlds and the positions that actors take there. For instance, engineers in the planners' world maintain their position by legitimising their technical and economic models of project evaluation. As a consequence, every actor participates in the decision-making process with his own view of the problem to be solved and how this should be done. Both representations are based on specific, quite individual, characteristics. Moreover, they may have little or nothing to do with each other (a very classical contradiction for instance is that between the objectives to ease traffic flow and to restrict road circulation). The complexity stems from the difficulty of making actors coming from different social worlds interact, so as to reach a collective way forward.

The entry of these new actors gives a hybrid character to the contents of transport policies. These can be analysed as a transition from a production logic of action – to build an infrastructure, to provide an equipment – to a problem logic of action – for example to assert the process of economic development (or growth). This transition (set up by actors involved in the decision-making process) constitutes a practical acknowledgement of the interaction between the output of public action (the infrastructure) and the outcome of public action (the development of territories). The consequences of governmental acts matter more than the acts themselves (Duran, 1999). Two consequences can be drawn from this statement:

- the first is that major transport infrastructure policy has to be considered in conjunction with land-use policy and environmental policy. Both policies emphasise the interconnection of social problems and the multiple outcomes of public measures;
- the second is that major transport infrastructure policy exists only in a practical sense, notably territorial: a public policy combines a process of defining contents and an implementation process in an interdependent way and through action (Padioleau, 1998). Activities generally involved in public policy, notably the formulation of a political problem and the implementation of a solution, are interrelated and not separated (Lascoumes, 1996).

We shall approach the fundamentals through three main issues that the concrete systems of action generally set up:

- the financing issue is a way to answer to the question: how to do the project?
- the territorialisation issue is a way to decide: where to do the project?
- the environment issue is a way to answer the question: what is the project for?

These three questions can be seen as a progression in terms of time and space: from the most traditional way of decision-making to the most recent; from the most closed to the most open process. Actually, all three occur simultaneously in current decision-making relating to major infrastructure projects and they all three have to be taken into consideration by the actors involved. But the third question sets forward the problem of the justification of the projects, and interrelated to it, the way through which projects could be justified. This issue characterises a double dimension of decision-making: that of its public nature (i.e. decisions have to be exposed publicly) and that of the setting-up of exchanges (with the public, and between different social worlds).

Based on the results of two research projects, this paper will review how the concrete systems of action have provided answers to these three questions (part I, the European TENASSESS research) and how actors appeal for some specific roles in order to manage and make decision-making public (part II, the French inter-ministerial PREDIT programme of research on transport). On this basis, we develop some conclusions with regard to public action, decision-making and the role of evaluation.

Circularity, Territorialisation, Public Accountability

The research project TENASSESS provided the opportunity to undertake an analysis of the decision-making process related to the Trans-European transport networks and the problems and conflicts they produce in different national contexts.[1] For each project, concrete systems of action were analysed by identifying the strategies implemented by the different categories of players – mainly planners (administration), decision-makers (at national and local levels), economic interests, associations – and the issues that made them interact concerning the project's definition. The comparison of the different case studies highlighted three main common issues to each decision-making process.

How to do: Partnership, Feasibility and Circularity of Public Action

Title XII of the Treaty of Maastricht (1993) and the Christophersen list (1995),[2] placed the transport Trans-European networks on the European political agenda. The infrastructure projects were considered as one solution to the problem of integration and the constitution of a European space. To be part of the list called for a specific international dimension and the metaphor of the 'key links' helped legitimise the TEN – all the more as they were named 'missing links'.

The metaphor can be traced to the Round Table of Industrialists, a European lobbying organisation that acted to promote the programme of transport infrastructure investments finally inscribed in the Maastricht Treaty under the TEN acronym. This metaphor was used to title one of its reports (1984). Two years later, the organisation persisted with a new report titled *The Missing Networks*. As a specific political entrepreneurship implemented by the members of the Round Table and supported by Jacques Delors while he was the Commission's president (European Commission, 1993), the TEN became a solution (among others) to the constitution of a European space because the terms of the problem and of the solution were vague[3] enough to make different actors act together in order to progress the TEN projects.

However, all the major transport infrastructure projects that constitute the TEN are costly and the actors involved in their decision-making processes see their financing as an important issue. The European Union has proposed the consideration of public-private partnerships as a solution (High-Level Group on PPP Financing, 1997). This solution acts as a feasibility criterion which has to be considered as a norm of action in the field of major infrastructure projects. We approach it as a way to consider the question of how to do the project. And the answer explicitly links feasibility and financing.

This criterion of feasibility and the actors' wish to comply with it (in order to keep the project on the European agenda) induces a specific characteristic of contemporary public action: that of circularity. Circularity of public action means that several actors set together both the problem to be solved and the method to solve it, the categories of actors who act collectively determining, in turn, the problem and its solution. In other words if the actors were different, the terms of the problem, of the solution and of the way to reach it would also be different. The use of the PPP implies that specific categories of actors are needed to act together.

In the case the of High Speed Train projects, the need to link financing and feasibility has persuaded those actors involved to propose phasing as a solution: the project is defined in several parts to be built separately or sequentially, each one corresponding to equal 'realistic' budgets. For instance, in the case of the Lyon-Turin TGV or the Aquitaine TGV, 7 billion FF (1998) were considered as 'realistic budgets': budgets that the actors (public ones – RFF the owner of the infrastructures, the SNCF and the local authorities) assessed as realistic with regard to their own financial resources but also with regard to the overall functions of the project. For instance, in the Lyon-Turin case, the split corresponded to such parts that each actor could consider that his own interest had been taken into account, independently of the priority to be set between each part. This process is itself characterised by the circularity of public action.

To belong to the TEN required evidence that the project under consideration possessed a European dimension and as such supported the implementation of a European consistent space. This criterion, demanded by the European decision-makers, was in turn a way to legitimate the European level of policy formulation. This is the reason why all the TEN projects had an international dimension. The international part of some TEN projects was used as a basic opportunity to implement the European PPP criterion in a circular logic of action. Whether for the Øresund Fixed Link or for the Lyon-Turin and Barcelona-Montpellier TGV projects, the international part was identified within the whole project as a specific part involving specific actors and a specific way to manage it. The basis for identification was the operation of the international part: partnerships induce operation and resources earned from it to be shared between partners.

For the TEN projects, the financial partners are the public authorities (central and local), the transport companies and specific partners interested in the operation of the international part. In the Lyon-Turin case, for instance, the Mont-Blanc and Fréjus tunnel operators expressed such an interest. The interested parties are themselves determined by the feasibility criterion: such a criterion means explicitly that the international parts will be operated through concession systems (in the tunnels and TGV cases, for instance). Therefore, the search for PPP put forward a specific way to elaborate the infrastructure projects: to link financing and pricing principles. This interdependency can be explained by the concession system to be envisaged and by the category of actors to be involved (namely the operators).

In order to organise and to stabilise the relationships between partners, specific structures and entities were set up: the Scandinavian Link Consortium and the Øresund Consortium in the Øresund Fixed Link case; a European Economic Interest

Group (EEIG) in the Lyon-Turin and Barcelona-Montpellier High-Speed Train cases. We call these specific entities *ad hoc* structures: they have been set up by actors that consider themselves partners and who collectively mobilise to reach a common goal (to get the infrastructure built) and to gain from the association. The main forms of knowledge linked are technical, economic and financial. All benefit from the high expertise capacity of the central administration and financial sector.

Hence also the circularity of action since the problem, the solution, the actors and the opportunity are closely interrelated. It is not easy to determine which of these elements is the explanatory one: did the solution (PPP) come after the problem (to build the infrastructure), to then determine the categories of actors to become partners, or did the causal pathway run the other way round? The most probable situation was that some actors were more active than others, thus successful in placing one or a set of specific projects on the agenda in turn to create an *ad hoc* structure as a way to integrate the feasibility criterion, the categories of actors potentially able to implement it, and their key interests. Such a structure is furthermore a way to stabilise partnerships and the priority of the project over time.

This partnership approach means that the interests of the participants have to be confronted and that a negotiation process characterises the decision-making, the results of which must be a sort of a compromise. The problem for managing public action then becomes that of regulation, in order to achieve compatibility between issues and consistency of the different strategies (Duran, 1999). Among the partners, the local authorities find themselves in a special position since they also represent the problem of where to do the project, which requires another kind of assessment.

Where to do (What): The Emergence of a Territorial Knowledge

As transport infrastructure projects have to be integrated into specific spatial areas, the local authorities in charge of those territories come to play a special role. We coined the term territorialisation to describe the process of spatial integration of an infrastructure project at regional or local level (cf. Offner and Pumain, 1996; Duran and Thoenig, 1996).

Territorialisation comprises two moves: a 'bottom-up' process which characterises the way in which local authorities use a major transport infrastructure project as a means to construct a new legitimacy for themselves as public actors, and the activities they deploy to take a more direct part in the decision-making process; a 'top-down' process which relates to the wish of the state (central government) to bind several public actors in the implementation process with the explicit objective of sharing the responsibility and risk of a public decision.

Territorialisation represents the concrete test with regard to the project's implementation potential. The particularity of this test derives from the part played by local authorities and consists in putting forward the evaluation of the outcomes of the proposed output, insofar as the local authorities that are in charge of the welfare of specific territories act on behalf of their communities. Territories to be transversed by the project expect to gain some impact (mainly economic and also symbolic) from hosting the infrastructure: the co-ordination of actions in the field of transport and actions in the field of local (economic) development is constructed as proof of the participation of local authorities in the decision-making. The extension of the Northern TGV in Belgium exemplifies this process with the occurrence of two possible extremes as a result of the involvement of two separate local authorities (Anvers and Liège), which asked for two different kinds of projects (one for the construction of a new line, the other for the improvement of the existing line). In the Barcelona-Montpellier TGV case, the project represented the opportunity for Catalunya to elaborate a whole new transport system and related activities by supporting a mixed railway project (passengers and freight) that would set the conditions for developing the seaport of Barcelona and airport activities. The Lyon-Turin territorialisation process resulted in a complication of the project that would serve jointly international, national and local interests, as well as freight and passenger services. Owing to this process, French TGV projects are nowadays elaborated by thinking simultaneously about the international, the national and the local transport services.

The main issue within the territorialisation process is that of the relationships between central and local governments, since central governments remain responsible for the decision to build major infrastructures and for the greatest part of local budgets. For the local governments, participation in the decision-making process is often used to build up their autonomy *vis-à-vis* central governments. To gain more autonomy, they act by the way of their competencies, generally concerning the fields of local economic development, quality of life, land-use, and the organisation of local public transports. Therefore, the local authorities set directly the political agenda in terms of the consequences of the transport project for the territory they are responsible for, and they do this by stating the interdependency of social phenomena and the question of the consistency between actions in different fields.

The way the local authorities integrate their agenda into the project depends on the structure of power and on the framework of the relationships between central and local governments. The main difference comes from the mediation system between the two levels of government, whether direct or through the parliament. In case of direct mediation (in France, for instance, owing to the plurality of mandates), the problem is mainly set in terms of transport services at different geographic scales and relative to one mode (rail sector). In case of partisan or parliamentary mediation (in Denmark or Austria, for instance), the transport problem takes a strategic dimension, namely that of considering the transport system as a whole (rail and road together) and of introducing pricing principles as a way to combine different sources of financing (road tolls).

To justify their viewpoint and make it acknowledged as a legitimate political problem partly solvable by the transport project, the local authorities mobilise the knowledge linked to their competencies, supported by their own technical services that have to be confronted with the methods of the project's evaluation as run by the central administrations. In all cases, territorialisation puts forward a new kind of expertise based on a territorial knowledge, namely that of the perceived territory of the living conditions, and how people evaluate them. Against the global European trend of decentralisation, this territorial evaluation means also public participation in decision-making. Citizens have interests in what happens to the territory or space where they live, its environmental quality as well as its accessibility. This is the reason why the decision-making process relating to major infrastructure projects could take a strategic and political dimension, perceptible through environmental evaluation and its underlying question of: a project, what for?

A Project, What For: The Political Dimension of Environmental Evaluation

Environmental appraisal is today part of all decision processes concerning transport infrastructures. In the majority of the cases, the concrete product of such an appraisal is an impact study. This is a specific study which assesses and takes into consideration the damage resulting from the construction of the infrastructure. Since major infrastructure projects are today confronted with many protests and mobilisation against them, no decision-maker can ignore the outcomes of the latter. Public authorities are now subject to a double agenda: to solve the problem but also to make the decisions acceptable, both to express a collective interest and to develop a public management method, to govern and to manage (Leca, 1997). Ecological values and environmental concerns set up a scene of action where this double agenda acquires concrete elements. This is the scene where the legitimacy of action and of infrastructure projects is directly questioned.

The positioning of the environmental appraisal procedure in the institutional framework and in the decision-making process shapes both the strategies of the actors involved and the solutions they reach. When the central administration keeps control of the evaluation, a juridical logic of action can be observed: the environmental appraisal is aimed at conforming to legal procedure in order to avoid later appeals. In other words, administration routine is made in anticipation and linearity. The appraisal is mainly technical, so is the problem and the solution (Eastern TGV case, for instance).

As ecological values assume more importance, institutional frameworks have acknowledged them as legitimate by setting up a specific central administration in charge of this field of action. Supported by associative protests against major infrastructure projects, environment ministries tend to play a bigger part in the decision-making process. Whether at the central or at local level, the administration in charge of transport has to be confronted with interests and logics of action from the environmental perspective.

This confrontation is not entirely new. For instance, in France, the Environment Ministry has institutionalised impact studies for major projects since 1976 and progressively the Transport Ministry has included environmental dimensions in its procedures ruling the major infrastructure projects' evaluation (Ollivier-Trigalo and Piechaczyk, 2001). When the planners in administration deal with the 'environmentalisation' of the projects' evaluation, they keep the technico-economic construction of their models and consider that they improve them by integrating more and more environmental dimensions. Like any actor, they translate the problem they have perceived according to their own way of thinking and acting, so that the changes in their tools would be directly usable by and within their routines. But planners now perceive that the conflicts due to major infrastructure projects' reflect other needs that cannot be accounted for by 'simply' rendering the models more and more sophisticated. Be it by elaborating strategic evaluation (Tomlinson and Fry, 2002) or by promoting *ex post* evaluation (Lauridsen and Larsen, 2002), the actors appear to pay more attention to the outcomes of public actions and the interests of citizens. In other words, the publication of projects induces another dimension to evaluation than the technical and economical ones.

When the institutional frameworks give importance to the mediation through political parties and parliaments, the confrontation between transport and environment fields goes through the notion of sustainability and its strong anchoring in political discourse and culture. This was the case for instance with the Brenner projects and the Øresund Fixed Link. There, the environmental issue assumed a strategic dimension by the way of considering a societal vision of sustainable transport which put forward the question of: 'which sort of society do we want to live in and leave to future generations?' This logic of action can be qualified as one which relates to the multiple dimensions of a social system.

Then several kinds of expertise are called for environmental evaluation (technical, economic, but also historic, cultural, social, territorial, etc.). No administration working according to technical and standardised categories of action can undertake, nor support such a multidimensional appraisal, since this would be a way to question its own limited viewpoint.

Of interest here is the experience of the regional council of Rhône-Alpes (an elect assembly) in setting up a system of counter-expertise. This system allowed environmental associations to ask for a counter-assessment of the SNCF file prepared for the implementation of the Lyon-Turin project. This counter-assessment was financed by the regional council. The justification of such an initiative was to state the inconsistency of public actions in the field of transport in Rhône-Alpes, with regard to the consequences for the region in several sectors (road and rail infrastructures, public transport).

The environmental agenda sets forward interlinked elements: the consequences of public action, multi-dimensional evaluation, the plurality of viewpoints. Interrelated to these characteristics, it raises the question of democratic participation in the decision-making process. To take into account the consequences of public actions involves the need to justify such actions to people living in the territories to be transversed. Therefore, the legitimacy of actions and political power depends also on the existence of a public space or scene where the justification of actions will be publicly accessible, implying that evaluation becomes an exercise in practical democracy, not merely a management tool.

Information, Relationships and Roles

The analysis so far shows that conflicts are intrinsic to the legitimacy of infrastructure projects, both in terms of their appropriateness and the way of deciding if they are appropriate. Such conflicts cannot be approached anymore in terms of opposition between private interests and public interest alone. At stake in the conflict is the sense of the policies that any infrastructure project concretises. In our democratic societies, no specific actor can be the unique owner of the legitimacy of public action. Each actor who is directly involved in such conflicts enters and takes part in the decision-making with his own sense of his action. The confrontation of those different senses is at the core of the legitimation of the projects.

Starting from this hypothesis, we have organised discussion groups in order to give practical and concrete elements to such confrontations (Rui, Ollivier-Trigalo and Fourniau, 2001). Based on the sociological intervention method, we set up groups gathering persons who have experiences of public debates, consultations or informal dialogues occasioned by major projects' conflicts and implementation procedures. Four categories of protagonists were considered: associations (split into two groups of discussion), contracting authorities (or planners: two groups), mediators (the persons in charge of public enquiries: one group), elected representatives (one group).[4]

The aim of this method was to make everyone exchange views on a common experience: to say what they do, why they do it and to give their reasons for this. The persons were thus asked to discuss their practices exercised during public debates, their expectations about such participation, their judgements and their experiences.

The results of these discussions allowed us to characterise the emerging forms of participation in decision-making. In effect, we chose the persons that composed the groups on the basis of previous monographs we had undertaken about several conflicts. Furthermore, we appealed to the individuals we had personally met and identified as particularly activist with regard to their participation to decision-making. Our objective was not to be representative of each group as a whole but to understand new practices. Since public participation is that dimension which effects most changes in evaluation methods, of interest here was to report what were the viewpoints of the main categories of protagonists expressed with regard to their participation in decision-making, their dissatisfactions and their claims (Ollivier-Trigalo, 2001). Those viewpoints can be split into three main categories: two relate to the controversial elements – namely, the information and the relationships; the last refers to the roles expected. It was amazing to observe how close the experiences of the different categories of protagonists were. This amazement was shared by us and the discussion participants.

The Inappropriateness of the Information

A project set under debate is first known by a dossier which gives it a concrete public reality. Most often, the information takes the form of a descriptive file that the contracting authority (the planner) elaborates and sets on the table. To take part in the debate relative to the project, it is essential to get access to this information, to read it, and to understand it in order to build an opinion and to have something to say about the project. But a number of points illustrate the drawbacks of the usual information and the difficulties of setting-up a satisfying debate on this basis.

In France, the legal procedures that organise the public participation determine more or less precisely the contents of the descriptive files according to the different stages that mark out the progression of the procedure: as the procedure goes on, the contents of the files will be more and more precise. Thus, for the first step, the project is described through its services' principles and its functions. For the last step (the public enquiry), the project is described by a socio-economic balance for the collectivity, a financial balance for the future operator, the technical studies of the route (located in a 300 meters zone) and an impact study (notably, an assessment of the noise nuisances). Whatever the information and the stage have been, the different protagonists unanimously agreed together to consider that this basis for a debate is unsatisfactory.

First of all, the access to information is not so easy and according to the situations, the protagonists feel that there is too much information (the files are judged voluminous, unreadable, incomprehensible) or that there is not enough information (some files are not publicly circulated but can be obtained by specific claiming or negotiations).

Secondly, the quality of information is often questioned. Those opposed to a project often question the correctness of the information. This judgement is nuanced depending on what category of information is under consideration. If it is about the general objectives of the project, the information is judged barren, not backed up, not open to discussion with the authorities: all the rhetoric based on key words such as economic growth and development, employment, social progress, opening up, is systematically denounced by the associations. If it is about the forecasts (traffic, rates of return), there is no end to the debate over their validity. If it is about the alternative routes, the character of alternative itself appears as doubtful (some alternatives are quite explicitly set as foils).

Thirdly, the nature of the information also induces hard criticisms: the problem here is that of the relevance of the information with regard to the matter of the debate. Often the protagonists appear to misunderstand each other, with the feeling that they do not talk about the same matter. The best example of such misunderstanding is the first stage of the procedure which indicates that the debate must be on the functions of the project but that any consideration about the possible routes has to be excluded (in order to be discussed at a further stage). In practice, this theoretical split in two debates cannot be sustained. What is at stake here is the problem to know what can be discussed: if the project is not precise enough, the debate finds no participants; if the project is precise, the debate turns out to be a dispute with no influence on the decision (the project seems already decided and defined).

Finally, information and debate appear to be constantly inadequate. What is noticeable is that this failure of appropriateness is felt at all stages of the procedure and by all the protagonists even if the inadequacies are not the same for every category of protagonists.

The Relationships' Spaces: The Taking into Account of the Arguments

A debate creates spaces for relationships. The judgements are not much more positive than for the information, apart for the mediators who are the protagonists that probably control the best those spaces. One difficulty arises from the fact that such experiences are relatively new for all. As a consequence, each one has his own opinion on how these meetings should be run, which refers to different visions of how to act together, collectively.

Through the procedures the contracting authorities have some room for manoeuvre to organise meetings with other protagonists involved in a project's decision-making. Mainly, the meetings are split into two great categories by all protagonists, dependant on their character – open or closed. Open means that the meeting is public and that everyone can attend it; closed means that the meeting is set up on the basis of individual invitations. These categories are built by the protagonists because they correspond to different kinds of relationships (in terms of form and contents).

Public or open meetings tend to at first be perceived negatively: the open meeting is not a debate – for instance, the politicians behaviour is often criticised since they use these meetings as a political tribune – but almost as a sort of a boxing ring. But, it is also recognised that a public meeting could have another utility than providing a forum for debate: its public character gives to the words said a status of words in front of witnesses.

The closed or restricted meetings allow more direct relationships. The contracting authorities are the first to support such meetings (with the mediators) because they use them as a way to make their project progress owing to a real dialogue between different protagonists, which is judged impossible during an open meeting. Two characteristics render such a dialogue possible: the restricted audience; a precise element or dimension of the project to be discussed (for instance, what kinds of protections could be added alongside the route in a specific zone). Yet another characteristic is sometimes criticised by the associations, namely that of their confidentiality: this characteristic appears as contrary to their collective basis especially when the authority that set up such a close meeting has the objective to end it with a commitment on the project.

Finally, the associations are the most disappointed with regard to the quality of the exchanges, notably with the contracting authorities, and are the most virulent with their judgements. This phenomenon is not very surprising: it gives a measure of their dissatisfied claims induced by the effort they make to participate in those exchanges. The same negative statement is elaborated by the contracting authorities that also feel dissension and lack of a common language, which makes the public debate a space that could appear as relatively empty. Actually, the protagonists do not stay with arms folded but take some initiatives to make decision-making more transparent. Those initiatives can be described in terms of the roles played by each.

The Roles in the Public Debate

On the public debate scene, each protagonist has experienced a role assigned by the institutional decision-making framework: that of roadside resident, planner or politician. We could split these roles in two categories: the ones who act and those asked for what they think of these acts. In the first category we find the contracting authorities and the mediators; in the second one, the public. But, among the protagonists, there are also the elected representatives who can be located on either side of that dividing line dependent on the situation.

In the current context where public decision faces legitimacy problems, the particularity of public debate is to reveal the failures of such definitions. The associations which come to the debate representing the local or wider public find that it is difficult to make their voices heard in such a forum and that even if they succeed, little attention is paid to them. They often feel they are relegated to a place of secondary importance. The elected representatives are, on the one hand, specifically consulted because of their mandates, on the other they may take part in the same open spaces as the public. As a consequence, within the same space they have to combine their relationship with the state that promotes the project and their relationship with their constituency (that may be opposed to the project). They perceive their situation as shaky. The contracting authority is set on the front line, both because it is the one who talks the most about the project and because it is often the one who calls the meetings. The contracting authority appears to be particularly exposed in and by the public debate. Alone the mediators do not feel threatened. This is partly explained by their singular place: an arbiter outside the conflict, and yet central to the decision-making.

Those assigned a dissatisfying status search for ways to improve their position. The protagonists act according to their own motivations and, by doing so, each one claims another role: citizen, political entrepreneur, leader of the debate, the one who delivers the needs of the public, etc. It appears then that the public debate introduces a new scene for legitimating decisions where quasi ideal-type roles not clearly played are appealed for. Based on the accounts told by the participants to our groups, six roles were identified.

- *The decision-maker* who seems to be totally absent from the debate: the protagonists never find a political representative able to demonstrate that a decision has been taken with regard to the project under discussion.
- *The project's promoter*. Distinct from the decision-maker, the project's promoter is the protagonist who explicitly appears with the will to make the project progress by defending it so that a possible solution would be set according to the local situation where it should be implemented. This role is both political and technical.
- *The expert* whose main role should be to explain the object of debate because of the high technico-economic dimension of major projects. The expert may give a technical and economic advice on the project; he or she may help the judgement of each on the project. The objective here is to allow access to debate.
- *The public* in the sense that they should express their viewpoint that they have developed owing to the debate. Often the public feels to be only an alibi to the debate or to the project, whereas they wish to play a counter-power part.
- *The debate organiser* is the one who should ensure that a debate occurs between protagonists: he has to instil confidence between participants, by providing a matter for discussion, by mobilising the public or by leading the debate.
- *The outsider* plays an arbiter and a peacemaker role, in order to re-launch the debate: the whole process being a situation of conflict by nature, there is a need for another party who warrants the respect of the game rules and who is not one of the protagonists.

Conclusion – Coordination, Cognition and Political Dimensions

The analyses of decision-making relative to major infrastructure projects exemplify some contemporary characteristics of public action. The identification of the different categories of actors involved and of their interests, viewpoints and strategies, confirmed their diversity in terms of practices, categories of knowledge and of action, and in terms of social sectors to which they belong.

The financing scene of action and its (part) solution through PPP shows how public and private practices and sectors try to build together a common political problem and its solution – very roughly, we could say that the public part had to integrate private considerations such as financial profitability, and that the private part had to take care of public interests such as safety, legal rules or public acceptability. The other scenes of action also showed that a contemporary public action issue would be that of the relationship-setting between different social worlds aiming at constructing a common political problem to be solved, notably by involving public authorities that are specific actors in collective and organised action.

The problem for decision-makers then is that of managing these new issues with regard to these conditions of action. These conditions for governing appeal for a specific competency in the field of understanding, relating and adjusting data coming from different social sectors. This decision-making framework gives high importance to the role of the intermediary, the one who ensures both the relationship-setting between actors and the providing of common meanings (Lascoumes, 1996).

Public action therefore demonstrates a need for a co-ordination role in order to set up a space for relationships between knowledge and expertise – i.e. to organise the underlying cognitive activities included in any public policy elaboration. In the field of the TEN, different kinds of knowledge are emerging: technical, economic, financial but also social, cultural, historic and territorial. Besides, the openness of the decision-making processes towards citizens induces also the need to organise a public space of relationships between decision-makers, planners and population. In the field of the TEN, the administration (at central and local levels) that keeps for statutory reasons a central role of mediation should take into consideration the political dimension of its activities that this openness would induce (Rui, 2001).

Notes

1 The whole problematic of the research project was initiated by ICCR (Interdisciplinary Centre for Comparative Research in the Social Sciences). The ICCR's team who got involved in the *TENASSESS* project was mainly composed by sociologists and political scientists. To comply with the wishes of the call for tenders proposed in the framework of the DGVII' 4th FPRD, ICCR set up a consortium associating 14 partners and co-partners coming from other Member States. The other partners were: Halcrow Fox (GB), INRETS (F), PLANCO Consulting GmbH (G), SYSTEMA (Gr.), IVTB (D), ERRI (GB), TRT (I). The co-partners were: UKO, NEA, NEI, LESEC (Esp.), ICCR-London, UWCC. The project started in May 1996 and ended in April 1999.
Ten case studies were carried out:
-The Øresund Link between Denmark and Sweden, one of the few successfully implemented TEN priority projects;
-The Brenner axis, likewise a major focus of the TEN – the part of the axis identified as priority, namely the tunnel, has still to be built;
-The Betuwe railway line between Germany and the Netherlands which is of potential

significance for the Northern European ports in relation to the opening to the East;
-The Twente Central Canal Connection in connection with the Betuwe Line;
-The Inter-island passenger transport system, a programme approved under INTERREG which aims at establishing a helicopter network to increase the accessibility of the Greek islands;
-The Barcelona-Montpellier Link, another TEN project and a major TGV project for Spain, and in particular Catalonia, which aims to effect a fast train connection between Spain and France with important implications for the port of Barcelona;
-The Lyon-Turin link Transalpine Railway Connection a project made up of numerous split projects covering the whole regional network;
-The Eastern TGV towards Strasbourg and Luxembourg, a project which for the first time raised doubts about the high-speed rail project;
-The TGV PBKAL Brussels-Amsterdam/Köln the implementation of which represents a serious re-formulation of the high-speed rail towards an integration of regional concerns;
-The Skaramanga Interchange out of Athens, a project of national relevance but displaying conflicts typical of major infrastructure projects, albeit experienced for the first time in the Greek context.
-All the case studies were included in TENASSESS, Deliverable R(3) – Technical Annex: Case Studies, 1997.

2 Group of personal representatives of the Head of State of Government.

3 The terms were vague (cohesion and sustainability, if we update the terms) and based on a symbolic rethoric (the structuring effects of transport infrastructures on economic development) which acted as a myth (see Peters, in this book).

4 Two researchers conducted the discussions each time. We met each group twice: the first time during one day long and three different sessions; the second time, two weeks later, during one session. The first three sessions were split as following: the two first sessions were dedicated to discussions between the group's members and another kind of decision-making protagonist (for instance, one group of associations faced a planner and a representative of a national environmental association); the third one was closed and gave the opportunity to the group's members to exchange experiences. Two weeks later, the researchers communicated their very first analysis of the first discussions and led a new debate with the group.

References

Crozier, M. and Friedberg, E. (1977), *L'Acteur et le Système*, Éditions du Seuil, Paris.

Duran, P. (1999), 'Penser l'Action Publique', L.G.D.J., *Série Politique, Droit et Société*, Maison des Sciences de l'Homme, Paris, p.212.

Duran, P. and Thoenig, J. Cl. (1996), 'L'État et la Gestion Publique Territoriale', in *Revue Française de Science Politique*, Vol. 46, No. 4, Paris, pp.580–623.

European Commission (1993), White Paper on 'Growth, Competitiveness and Employement'.

Friedberg, E. (1993), *Le Pouvoir et la Règle, Dynamiques de l'Action Organisée*, Seuil, Paris.

High-Level Group (1997), on 'Public-Private Partnership Financing of TEN Projects', Final Report, VII/321/97, European Commission-DGVII, Brussels.

Larsen, O. (2002), 'Norwegian Urban road tolling: What Role for Evaluation?', in Alan Pearman and *alii*, *Transport Projects, Programmes and Policies: Evaluation Needs and Capabilities*.

Lascoumes, P. (1996), 'Rendre Gouvernable: de la 'Traduction' au 'Transcodage'. l'Analyse des Processus de Changement dans les Réseaux d'Action Publique', in *La Gouvernabilité*, CURAPP, PUF, pp.325–338.
Lauridsen, H. (2002), 'Strategic Transport Planning and Evaluation: the Scandinavian Experience', in Alan Pearman and *alii*, *Transport Projects, Programmes and Policies: Evaluation Needs and Capabilities*.
Leca, J. (1997), 'Le Gouvernement en Europe, un Gouvernement Européen?', in *Politiques et Management Public*, Vol. 15, No. 1.
Offner, J. M. and Pumain, D., (Dir.) (1996), 'Réseaux et Territoire – Significations Croisées', *Éditions de l'Aube, coll. Territoire*, La Tour d'Aigues, p.281.
Ollivier-Trigalo, M. (2001), 'Expériences du Débat Public, Insatisfactions et Revendications', in Rui, S., Ollivier-Trigalo M. and Fourniau, J.M., 'Débattre ou Négocier l'Utilité Publique?', *l'Expérience de la Mise en Discussion Publique des Projets: Ateliers de Bilan du Débat Public*, PREDIT 1996–2000–DDT, Rapport de Recherche No. 240, INRETS, pp.39–98.
Ollivier-Trigalo, M. (2001), 'The Implementation of Major Infrastructure Projects. Conflicts and Co-ordination', in *Transport Policy and Research: What Future?*, Avebury, pp.19–45.
Ollivier-Trigalo, M. and Piechaczyk, X. (2001), 'Evaluer, Débattre ou Négocier l'Utilité Publique?', *Le Débat Public en Amont des Grands Projets d'Aménagement: un Thème pour une Communauté d'Idées*, PREDIT 1996–2000–DTT, Rapport de Recherche, No. 233, INRETS, p.304.
Padioleau, J.G. (1982), 'L'État au Concret', *PUF., coll. sociologies*, Paris, p.222.
Padioleau, J.G. (1998), 'Prospective de l'Aménagement du Territoire: Refondations Liminaires de l'Action Publique Conventionnelle', Université de Paris-Dauphine, GEMAS/MSH, Décembre, p.25.
Peters, D. (2002), 'Old Myths and New Realities of Transport Corridor Assessment: Implications for EU Interventions in Central Europe', in Alan Pearman and *alii*, *Transport Projects, Programmes and Policies: Evaluation Needs and Capabilities*.
Rui, S. (2001), 'L'Expérience Démocratique, l'Implication des Citoyens dans les Procédures de Débat Public Autour des Grands Projets d'Aménagement', Thèse pour l'Obtention du Doctorat en Sociologie, sous la Direction de François Dubet, Département de Sociologie, Université de Bordeaux II, Présentée et Soutenue Publiquement le 5 Janvier.
Rui, S., Ollivier-Trigalo, M. and Fourniau, J. M. (2001), 'Evaluer, Débattre ou Négocier l'Utilité Publique?', *l'Expérience de la Mise en Discussion Publique des Projets', Ateliers de Bilan du Débat Public*, PREDIT 1996–2000–DDT, Rapport de Recherche No. 240, INRETS, pp.358.
TENASSESS (1997), 'Deliverable R(3)' – Technical Annex: Case Studies.
Thoenig, J. Cl. (1985), 'L'Analyse des Politiques Publiques', in Grawitz, M. et Leca, J., (Dir.), *Traité de Science Politique*, Tome 4, PUF., Paris, pp.1–60.
Tomlinson, P. and Fry, C. (2002), 'Strategic Environmental Assessment and its Relationship to Projects', in Alan Pearman and *alii*, *Transport Projects, Programmes and Policies: Evaluation Needs and Capabilities*.

Chapter 12

Involving Stakeholders in the Evaluation of Transport Pricing

José M. Viegas and Rosário Macário

Urban Public Transport (UPT) in Europe is considered an indispensable element to support economic and social activities in modern cities, and this is probably the main reason why this sector is so politically sensitive and has been subject to state intervention all along its history, both through regulation and subsidisation.

As cities grow and consumption demands become more complex, mobility becomes also an essential asset to undertake other economic and social activities. Traditionally, state interventions in Public Transport have been partly justified by equity considerations, namely to ensure that the transport network was available to all citizens and that no one should be deprived of its services by considerations of price. The interpretation of this goal, which is in itself still valid nowadays in the implicit concept of public service, led the authorities to increase the financing of Urban Transport through the use of administratively set prices, concessionary fares and subsidies to cover companies' deficits. The main factors leading to this essentially political attitude were the willingness to compensate the insufficiency of revenue caused by a loss of patronage in favour of private cars, together with the lack of freedom from the operators to establish competitive prices.

Along the years, evidence revealed that this was not an efficient way of intervention and allocation of public money, once all users benefited from the same (subsidised) fares independent from their income levels. Additionally, there is a growing awareness that, to achieve a sustainable balance between private and public means of mobility, pricing policies have to be able to send the correct signals in order to induce an adaptive behaviour from the users, which in turn will provide the system with a reliable feedback on the needs for further investment and expansion of transport facilities.

In line with this evolution the EU Common Transport Policy (CEC, 1992) highlights the link between the improvement of transport infrastructure and the accessibility of the regions. According to the main lines of this policy, citizens and enterprises should have access to a mobility level corresponding in quality and performance to their expectations and needs and at a reasonable cost.

This switch of perspective is also reflected in the aims of the *Citizens' Network* Green Paper (CEC, 1996), where the European Commission states the importance of assuring that the needs of the citizens are put at the centre of decisions about transport provision. The goal must be the achievement of networks of public passengers systems. Furthermore, Public Transport should ideally be a service open to all citizens in terms of accessibility to vehicles and infrastructure, affordability in terms of fare levels, and availability in terms of coverage of services.

In line with the previous political statements, in 1995 the European Commission launched another Green Paper under the title *Towards Fair and Efficient Pricing in Transport: Policy Options for Internalising the External Costs of Transport in the European Union* (CEC, 1995), advocating that:

- pricing should be seen as a complement of regulatory and other market policies;
- the main aim of a fairer and more efficient pricing policy is to use price signals to curb congestion, accidents and pollution;
- prices should reflect underlying scarcities to ensure sustainable transport;
- appropriate infrastructure charging is needed to mobilise private capital and relieve pressure on public budgets;
- the transport price structure should be: clear to transport users; differentiated across time, space and modes; non-discriminatory between modes and Member States.

More recently, a European Commission's White Paper on *Fair Payment for Infrastructure Use: a Phased Approach to a Common Transport Infrastructure Charging Framework in the EU* (CEC, 1998), though not directly applicable to Urban Transport, reinforces that:

- in future, charging systems should be based on the 'user pay' principle, i.e. marginal social cost basis;
- to achieve it a phasing system is proposed entailing:
- The adoption of a 'broadly compatible structure' in the main modes of transport until 2000, i.e. the first phase;
- a second phase dedicated to the harmonisation of the charging systems, with the implementation of a kilometre based charging system, differentiated on the basis of vehicle and geographical characteristics;
- finally, the third and last phase envisages the update of the implemented framework based on the experience gained in the previous phases.

From this evolution we can conclude that the transport policy goals can be grouped into three main categories, which reflect the fact that equity and efficiency are still the main criteria for policy appraisal (Gwilliam, 1987):

- allocative efficiency of resources within the transport sector and between this sector and other economic sectors;
- meeting individual requirements at minimum resource costs, that is market and productive efficiency objectives;
- equitable distribution of benefits and costs, that is equity objectives.

However, the definition of an Urban Transport policy is a complex issue since it is very much related with the specific characteristics of the local environment, as well as with the respective political options, which may change between localities within the same country, and even between neighbour communities served by the same transport system. The diversity of variables involved causes a wide diversity of approaches to Urban Transport policy that are reflected in the definition of a number of elements of the system, among which pricing and financing policies have a special role.

During the last two decades this sector experienced strong movements of change with three main goals: increasing its productive efficiency; reducing the gap between the price paid by the users and the real costs of providing the service; and reducing public expenditure in the sector by introducing new ways to involve private finance.

To solve these problems, new pricing and financing schemes have emerged all over Europe during the recent years, though only some of them have successfully survived the implementation process. These implementation difficulties often arise from the difficulty of identifying winners and losers and consequently devising effective schemes, and accompanying measures to transfer the gains of the former into compensation to the latter.

The success of pricing and financing schemes in Urban Transport Systems is also strongly dependent on the regulatory and organisational framework of the system, and on its potential to co-ordinate between the different policies with impact in mobility demand patterns (e.g. land-use, environment, etc.), as well as between the push and pull measures developed by different agents of the system – authorities and operators. It is often forgotten that one of the remote causes of the external costs caused by transport is the location of the economic and social activities, which create the mobility needs.

In addition, it is also the role of the transport pricing policy to contribute to the control of these external costs produced by the system, and this can only be achieved by the combined use of market based incentives together with control regulations. The former should persuade users to adapt their behaviour towards the policy aims, while the latter are mostly meant to restraint practices leading to the growth of external costs.

It is also worth to highlight at this stage that pricing and financing policies imply the use of different mechanisms at the same time, that is single measures are never fully effective if applied in an isolated way. The risks involved in the implementation of each measure, their synergetic potential, as well as the assurance that the different measures involved in one policy package do not produce contradictory effects, are important issues that have to be included in the concerns of the decision-makers when choosing the most appropriate policies for their local packaging.

Like in all other policy processes also here a policy cycle can be conceived to facilitate the identification of relationships with stakeholders in its different stages as well as the selection of the most adequate method to evaluate progress at each stage and obtain feed-back to finely tuned policy adjustments. Only this way evaluation can be seen as a self-learning tool for policy fitness.

Public Acceptability: An Interplay between Different Aspects

The changes in structure and dimension of the urban environment added to the congestion phenomenon, the scarcity of public money and, last but not least, a growing awareness of society about environmental problems, are among the main factors that have led to stronger demands for efficiency in transport systems, and consequently to the use of pricing policies as the main instrument to achieve that aim.

However, and despite a quite consensual recognition of all the advantages of this type of solution, there is a strong public opposition to the implementation of these schemes when they include charging for road use in urban areas, built upon the following main arguments:

- have to pay for what was previously free;
- excessive privilege accruing to the wealthier elements of society;
- no firm guarantees given for a fair and efficient application of revenues;
- threat to citizens' privacy.

The analysis of the opposing arguments to the transport pricing proposals included in the Commission's Papers[1] reveals a number of additional arguments that seem to be of high relevance in terms of public acceptability (PATS, 2000). These include:

- the equal and fair treatment of all transport modes and all sectors of the economy;
- the implications of transport pricing on European competitiveness and the social and economic environment;
- the qualification and quantification of transport costs;
- the adequacy of the pricing mechanism to create a significant modal shift from the road to more environmentally friendly modes;
- the availability of the technology for accurate transport charging;
- the use of revenues;
- fair and equal treatment of users;
- integration of transport pricing with other policies (e.g. urban planning/land use, regional policy, etc.).

These arguments show that the acceptability of transport pricing is as dependent on practical/functional issues, convictions and beliefs of the stakeholders as on its economic principles and foundations. From the literature review undertaken in PATS research (PATS, 2000) it clearly emerged that the perceptions and attitudes of a variety of stakeholders have a major part to play in gaining acceptability of pricing measures. These attitudes might differ between people, and between different groups of stakeholders, since the way a measure is perceived to impact on the daily lives of an individual might heighten or lower its acceptability. Whenever a pricing measure is perceived to force behaviours not in compliance with usual habits there is a strong potential to develop resisting patterns, unless advantages of those changes are made very obvious (Lippmann, 1997). If resistance is the option then the pricing measure can have unintended, and sometimes uncontrollable, collateral consequences and even totally fail to meet its objectives.

When implementing pricing measures it is therefore fundamental to consider the perceptions and attitudes of the general public and stakeholders. It would appear from the literature, particularly from social psychology, that one method to overcome potential opposition might be to develop marketing and publicity strategies, as well as information campaigns to be implemented together with the technical measures. These marketing strategies could be targeted at different groups and different perceptions with an aim of influencing them in favour of the pricing measure, and they should be incorporated in a global participatory approach.

It is worth noting that not only the fairness of price determination, but also the magnitude of price has a major influence on its acceptability. Consequently, transport cost calculation, which is the basis of a fair and efficient price, is also one of the aspects to be considered as of utmost importance. However, the fields of transport costing and pricing (especially for externalities and infrastructure) seem to be a very sensitive one as several methods and procedures are followed throughout Europe, that are likely to produce significantly different results (UNITE, 2000). These methodological uncertainties and disputes have to be mitigated and the evaluation process should be made coherent and transparent. The recently concluded FISCUS (1998) and on-going UNITE (2000) research projects, also funded by the Commission, are expected to bring new insights into this discussion by proposing cost calculation methodology and the implementation of harmonised national transport accounts that will enable future comparison between countries and consequently between citizens (stakeholders) of different countries.

Sophisticated technical systems play also an important role in the enhancement of fairer and more efficient pricing. However, they may also impose constraints in terms of acceptability due to complexity, lack of transparency, potential for violation of privacy and mistakes, inconvenience imposed on the users, etc. This means that among all the factors that contribute to 'build' the acceptability of a new pricing policy, the technical features of the system are extremely important as they are normally in the front line of users contact.

Finally, the careful choice of which authority is to regulate, implement and administer any pricing measure, and the legitimacy that an agency has in the eyes of the stakeholders may also improve acceptability. Revenues from pricing schemes attributed to the local authority may enhance the willingness of people to pay, as they expect a direct advantage from their application. Also, the trustworthiness of the local administration may be superior to that of the central government as decision-makers are closer to the field and can also be perceived as more familiar with stakeholders' interests. Different authorities have also different potentials to integrate a pricing measure with other policies to reduce congestion, accidents and pollution as well as to co-ordinate between different agents, which again constitute important elements in terms of acceptability.

Public acceptability is thus a complex problem as it requires the joint consideration of a number of scientific areas, in particular economic, social, technological, legal and even managerial aspects. At the top of all these aspects one can still add the fact that the analysis of acceptability implies the observation and control of a dynamic process of change where in one hand we have the individual interests of each citizen and its perception of advantage and disadvantage, resulting from the implementation of any pricing policy, and on the other hand there are still the organised interest groups who besides the weight of their public reactions have their lobbying power to be considered.

Equity: A Central Concept for Acceptability

Acceptability results from an interaction between political effectiveness, here understood as the capacity to accomplish the proposed objectives, economic efficiency in production and consumption, equity and social fairness and feasibility of implementation.

This multi-dimensional characteristic of acceptability should be considered from both the static and the dynamic perspective. That is, while introducing a pricing policy we have to consider not only the absolute effect on each stakeholder personally but also the marginal effects on the social and economic status of those persons at the moment of implementation and over time. That is, the impact of change must be assessed in short and long term and monitored over the lifetime of any pricing measure.

Efficiency and equity are thus central concepts for pricing transport systems. The first justifies the entrepreneurial attitude of decision-makers when defining and implementing pricing measures, the second covers an important part of what can be considered as the constraints on implementation.

Equity is a concept which has gained dominance in the industrialised societies, often being preferred as a distributional principle to govern economic relations or relations where an exchange of goods or contributions is the main concern (Schade and Schlag, 2000), such as the case of all public services. Some authors argue that educating people to be equity-minded is a determining contributor to the growth of performance oriented societies. The concept is born in social psychology and is seen as a motivational theory since perceived inequity may lead to feelings of guilt (when favoured) or anger (if others are perceived as favoured) often leading to reactance, the degree of which mostly depends on personal goals and style. In practical terms, equity is often understood as distributive justice entailing three possible main (alternative, frequently contradictory) rules for a just distribution: equality, meaning that everyone gets the same share of what is at stake; equity (*strictu sensu*), meaning similar personal input-output ratios between costs or contributions and accrued benefits; distribution based on needs or requirements, meaning everyone receives according to his needs or requirements.[2]

These distributive justice principles were translated in PATS[3] research along the following equity dimensions:

- territorial equity (or aggregate equity) – which corresponds to the principle of equality of opportunities and avoidance of social exclusion, and implies the consideration of the right to free mobility of people and goods. This right has a double effect when translated to the transport pricing domain: on one hand its preservation imposes limits to increases in transport pricing, on the other

hand that freedom has to be contained within the limits of the general interest of society;
- horizontal equity (or procedural) – which corresponds to procedural fairness by dealing differently with different circumstances, that is differentiation, implying a better coverage of costs by users (i.e. user-pay principle). However, pricing changes imposed by this principle might impose significant losses to some actors, leading then to reactance;
- vertical equity (or end state) – which corresponds to the assumption that any pricing policy has an intrinsic capacity of creating losers, by worsening the situation of the least advantaged groups or the least best served areas, or simply by openly assuming that the improvement of those situations is not aimed at. Groups of losers are of course potential drivers of reactance unless their losses can be compensated.

Last but not least, longitudinal (or dynamic) equity which represents everybody's goal of suffering no decrease of previously available benefits (entitlements), and corresponds to one of the major difficulties in modern, ageing societies.

From the above it can be easily understood that the main elements in the analysis of stakeholders' degree of acceptability of transport pricing policies are:

- the identification of the various actors involved or affected in all stages of the decision making process;
- the perception of the specific interests of those actors analysed as individuals;
- the power of influence of the different groups of actors and their reactance potential;
- and finally, the relevant aspects that can enhance the acceptance of the different pricing instruments and policies.

Finally, it is worth referring to the main practical aspects associated with the implementation stage, that should be carefully considered in the design of the measures and packages:

- functional interaction between institutions involved in the decision process, implementation and management of transport pricing and financing policies, as well as between these and other governmental areas;
- technological solutions necessary for implementation purposes, respective impact in citizens privacy, easiness to use and control, etc.;
- legal and regulatory framework in order to clarify the role of the different public and private intervening institutions.

Policy Cycle and the Use of Evaluation Frameworks

For many years the implementation of pricing policies followed a logically structured cycle of allocative decisions, based on a rational approach, going through the following steps:

- specify objectives;
- develop alternatives by which the objectives may be accomplished;
- evaluate the consequences of each alternative;
- chose the action that maximises net benefit according to decision-makers criteria and valuations.

This approach assumes a unitary decision-making, or considers a group acting as a unit, and ignores situations of conflict, which arise whenever social activities are at stake and the different interests of societal groups are confronted. However, policy definition is a much more demanding process as it requires the ability to define a problem according to different possible perspectives, to draw arguments from a diversity of sources, to adapt the argument to the audience and finally to educate public opinion in order to achieve consensus around policy objectives and design (Majone, 1989). That is to say that policy definition requires understanding the diversity of perspectives represented by the stakeholders, the interests they represent and the interdependence those interests impose in the different stages of policy definition.

Despite the criticism surrounding the staged approach to the policy process, that is the policy cycles, this is a model for policy analysis not devoid of advantages, in particular for its ability to provide a framework of thought which represents a simplification of reality reducing its complexity to a manageable dimension, and perfectly suitable for the case of transport pricing decision-making process. Notwithstanding, the argument of many authors can be easily accepted that the real world is far more complicated and not composed of tidy, net steps, phases or cycles (Parsons, 1995).

Less consensual are the criticisms that the model fails to consider interaction between actors, in particular between stakeholders and the different stages of policy life-cycle. When applied to transport pricing that cycle can be considered as having the following stages, where interaction with stakeholders can be reflected:

Stage 1. Definition of problems and objectives:

- perception of problems and dynamics of environment;
- definition of objectives, including gaining acceptance from stakeholders in setting the goals based on problem perception.

Stage 2. Policy design:

- identification of alternative solutions/responses to the problem:
- planning concepts/ future scenarios, including determination of favourable and unfavourable patterns;
- selection of policy instruments, including gaining acceptance from stakeholders for the selection of instruments which can reach the preferred defined patterns, assuming that each policy/instrument has its own goals and effects;
- assessment of expected impacts and stakeholders reactions (with possible feed-back to earlier items in this step).

Stage 3. Policy implementation:

- deployment of policy instruments including detailed design of legal, organisational, financial and operational instruments for the effective implementation of the policy;
- identifying and bringing together 'implementers', that is all individual and institutions involved in the set up of policy organisation, including gaining acceptance from stake-holders for that organisation;
- definition of an implementation plan, setting up resources, time and space scales.

Stage 4. Policy evaluation:

- checking stakeholders reactions;
- monitoring and evaluation of implementation;
- evaluation of policy impacts, side effects and overall outcome, by measuring the impact and policy success;
- if necessary return to stage 2 for fine-tuning (feed-back mechanism) and reconsideration of initial stages.

This staged approach is, however, not without criticism as, although considering the interaction with stakeholders, it oversimplifies the possible involvement of multiple levels of government in the process which can result in the existence of interacting cycles to consider the negotiation and bargaining process between those organisations. In addition, this approach is unable to reflect the political motivation to move from one stage to the next, nor is it able to assess any potential negative political balance that might lead to policy disruption in the middle of the process. Reflection on these criticisms will certainly lead us to the conclusion that the best analytical model will be a multi-framed one, though its main disadvantage will be the degree of complexity and the strong growth of interacting cycles that will considerably reduce our capacity of understanding the decision process entailed in each of its steps.

With respect to evaluation, we refrain to Parsons' (2000) concept that evaluation is fundamentally a process of valuing and that different frameworks of valuing inevitably generate different ways of thinking about the problem. So, evaluation is focused on the task of integrating knowledge (i.e. designing methods, processes and institutions which can best serve to clarify competing arguments) aiming to answer the question – who gets what knowledge when and how, and, ultimately, whose values get to dominate.

A number of evaluation frameworks can be selected to accompany the staged process defined above, although it must be stressed that the selection of the framework must be in line with the objectives of the different stages, which is why no best evaluation framework seems to exist for the overall transport pricing policy cycle, but instead, a more specific framework should do the job in each stage.

A multi-framework method for evaluation will form a basis for having these tools as self-learning instruments for policy definition and fine-tuning. Evaluation as calculating and distributing costs and maximising benefits and social welfare can be used in stage 2, while evaluation as the measurement and monitoring of performance is to be preferred for stage 4.

Whenever stakeholders' reactions are to be assessed (in the different stages of the model) then a pragmatic interpretivist stream frame seems to be more effective, since it adopts evaluation as education and empowerment of all stakeholders through dialogue and deliberative democracy, rejecting the positivist and objective claims of other frames in favour of an argumentative approach. This multi-frame path applied to the definition of transport pricing policy definition should facilitate the final and aggregate policy measure of evaluating through price systems, that is evaluation as allowing markets to facilitate experimentalism, learning and self-organising in conditions of uncertainty and complexity, just like the real world.

While the traditional model of rational choice assumes that people attempt to optimise their decisions within given rules, and that the underlying principle of maximisation of expected utility only guarantees that the choice is consistent with the decision-makers valuation of the probabilities and utility of the consequences of the various alternatives at stake, Majone (1989) suggests that the policy process becomes much more understandable if one assumes that actors view the rules of the policy game as possible targets of political action, striving for changing those rules in their favour.

Being so, the assessment of stakeholders' reactions cannot possibly be considered only at the stage of policy implementation, on the contrary the potential winners and losers status changes with time. That is, as previously said, when introducing a pricing policy we have to consider not only the absolute effect over each individual stakeholder at a given moment but also the marginal effects on the social and economic status of those individuals over time, meaning that the impact of a change in the status of the receptor of the policy effects must be continuously assessed in the short and long term assuming a dynamic evolution of that status.

This perspective allocates a dominant importance to the argumentative capacity at all stages of the political cycle, and even to the public acceptability phenomenon itself, since as Dewey (1927) once observed, the most important thing about popular voting and majority rules is less the current outcome of the voter choice than the fact that the electoral process compels prior recourse to methods of discussion, consultation, persuasion and the resulting modification of views to accommodate the opinion of the minority.

Overall Assessment of Transport Pricing and Financing Policies

Developing an evaluation framework for the assessment of transport pricing policies raises a number of problems, in particular for an urban environment, related to the diversity of the underlying legal and regulatory frameworks, as well as with the political priorities that can be at stake in the place where those schemes/measures are to be implemented.

Such an assessment framework must be flexible enough so that it can be used at different levels of detail: from the stage where only a theoretical and abstract description of those schemes is available, as it will be the case when conceiving new pricing and financing schemes, up to cases where experience exists and data on specific indicators is available. To respect these requirements a holistic approach is needed, with recourse to fuzzy assessment when the level of abstraction is high.

From the discussion of the previous sections of this paper and the findings of the referenced research the following conclusions can be drawn regarding the main guidelines that should be followed in the construction of such an assessment framework:

- typical conflicts exist between efficiency and distributional issues. For this reason when choosing among different pricing and financing schemes political decision-makers have often to establish trade-offs between these two dimensions;
- a 'fair' pricing policy can only be determined through the analysis of the impacts on the different stakeholders groups, as each schemes creates winners and losers;
- the effectiveness of any pricing scheme can only be assessed if the use of the revenues is also included in the analysis, as this represents a major factor in the process of financing system upgrading and in public acceptability;
- price and quality are closely related aspects. Investment and pricing decisions should not be taken independently from decisions on elements related to the quality of the service, such as network design and performance of the whole transport system, since price is a factor which influences customers expectations and perceptions regarding the quality of the service;
- for transport pricing to be an effective tool to curb congestion and other external costs, an integrated approach between the different policies involved in the urban environment is needed;
- evidence shows that some theoretically economic sound schemes have failed to survive the implementation stage due to several reasons, such as: cases where they represent a threat to citizens' privacy; practicability; inadequate organisational framework to ensure enforcement, etc.

These conclusions, together with the consideration of experiences reported in the literature, suggest the need to consider three different levels of assessment for each policy package (FISCUS,1998):

- aggregate policy level – addressing the economic and social effects concerning: economic efficiency in consumption and production, cost coverage of public expenditure, distributional effects (as seen from the 'top') and ecological sustainability;
- stakeholders' level – addressing the social acceptability effects, in particular the positive and/or negative impact perceived by the groups directly or indirectly affected by the system;
- practical feasibility level – addressing the issues related to the implementation stage, such as: legal, regulatory, technical and managerial issues.

The different characteristics of the schemes to be assessed call for the utilisation of different methods in the three assessment levels.

The first level of assessment, the aggregate policy level, deals with aggregate economic and social effects, and suggests the use not only of a social cost-benefit analysis but also the disaggregation into a hierarchical tree of criteria.

For the second level of assessment, concerning stakeholders' reactions, we should remember that the financing of urban mobility concerns not only the agents who directly or indirectly pay for the provision of infrastructure and services, but also the ones who bear its positive and negative impacts. Since politics in a democratic state must be carried out with permanent evaluation of levels of support and antagonism to the policies put forward, this evaluation is essential to ensure that the results remain realistic.

Finally the third level is dedicated to the practical feasibility of implementation of the schemes. This covers several steps to be carried out in sequence, considering the local environment where the scheme is to be implemented:

- legal, technical, managerial feasibility;
- necessary accompanying measures (e.g. land-use, fiscal, regulatory adjustments);
- costs of change against the benefits of implementation of the new scheme;
- brief assessment of the main risk factors against a successful implementation;
- global assessment of practical feasibility based on the conclusions from the previous steps.

The global assessment framework presented herewith aims to reflect not only the complexity of the urban environment, and respective organisation of the transport system, but also the fact that the same pricing and financing measure and/or policy package might produce different results in terms of efficiency and effectiveness, given the different underlying conditions for its implementation and the political goals strategically defined.

Conclusions

The work presented above aims to show the complexity entailed in the public acceptability phenomenon and to emphasise its multi-disciplinary character as well as its growing dominance as a critical factor for the success of transport pricing policies.

We have shown that it is possible, and helpful for policy analysis purposes, to conceive a policy cycle entailing the dynamics of the relation with stakeholders and the selection of adequate evaluation frameworks to monitor progress at each stage and to obtain feedback to fine tune policy adjustments and build a self-learning tool for policy fitness.

The importance of argumentative models was highlighted as an opposition to the traditional rational decision model, which to our understanding does not cope with the wide range of elements and dynamics involved in acceptability issues.

Finally, an overall assessment framework for transport pricing was presented showing that acceptability is one of the three main building blocks for successful transport pricing policies.

Notes

1 Analysis of reaction letters from stakeholders (individuals and organisations) done in the framework of the research project PATS.
2 Schade *et al.*, in AFFORD Deliverable 2C to the EC.
3 PATS research, Deliverable 2 to the EC.

References

CEC (1992), *The Future Development of the Common Transport Policy – a Global Approach to the Construction of a Community Framework for Sustainable Mobility*, COM (92), p.494.

CEC (1996), *The Citizens' Network – Fulfilling the Potential of Public Passenger Transport in Europe*, Office for Official Publications of the European Communities, ISBN 92–827–5812–5.

CEC (1995), *Towards Fair and Efficient Pricing in Transport – Policy Options for Internalising the External Costs of Transport in the European Union*, COM (95), p.691, Green Paper.

CEC (1998), *Fair Payment for Infrastructure Use: A Phased Approach to a Common Transport Infrastructure Charging Framework in the EU*, Brussels, CEC.

Dewey, J. (1989 [1927]), 'The Public and its Problems', in Majone G. (ed.), *Evidence, Arguments and Persuasion in the Working Process*, Yale University Press.

FISCUS (1998), *Cost Evaluation and Financing Schemes for Urban Transport Systems*, Transport Research Fourth Framework Programme, Urban Transport, Project co-ordinator: TIS, Transportes Inovação e Sistemas a.c.e., Lisbon, Portugal.

Gwilliam, K.M. (1987), 'Market Failures, Subsidy and Welfare Maximisation', in Glaister S. (ed.), *Transport Subsidy*, Special Issue *Policy Journal*.

Lippmann, W. (1997 [1922]), *Public Opinion*, Free Press Paperbacks, New York.

Majone, G. (1989), *Evidence, Arguments and Persuasion in the Policy Process*, Yale University Press.

Parsons, W. (1995), *Public Policy – An Introduction to the Theory and Practice of Policy Analysis*, Edward Elgar, Aldershot, UK.

Parsons, W. (2000), 'Analytical Frameworks for Policy and Project Evaluation: From Welfare Economics and Public Choice to Management Approaches', Proceedings of TRANS-TALK Workshop 1 'Policy and Project Evaluation: Context, Theory and Methods', 29th-31st May 2000, Brussels, ICCR.

PATS (2000), *Pricing Acceptability in the Transport Systems*, Transport Research Fourth Framework Programme, Urban Transport, Project co-ordinator: TIS.PT, Consultores em Transportes Inovação e Sistemas a.c.e., Lisbon, Portugal.

PETS (1998) *Pricing European Transport Systems*, Transport Research Fourth Framework Programme, Strategic, Project co-ordinator: ITS, Institute for Transport Studies, University of Leeds, Great Britain.

Schade, J. and Schlag, B. *et al.* (2000), 'Acceptability of Marginal Cost Road Pricing', AFFORD, Deliverable 2C to the European Commission.

UNITE (2000), 'Unification of Accounts and Marginal Costs for Transport Efficiency', *Transport Research Fifth Framework Programme*, Project co-ordinator: ITS, Institute for Transport Studies, University of Leeds, Great Britain.

Viegas, J. and Macário, R. (1998), 'Pricing and Financing Schemes in the different regulatory and organisational frameworks for Urban Transport System in Europe', TERA Conference, Milan.

Chapter 13

Accessibility Analysis Concepts and their Application to Transport Policy, Programme and Project Evaluation

Derek Halden

Introduction

The potential role for accessibility measuring techniques as part of transport evaluation in the EU, depends partly upon the evolving policy and administrative approach to transport. In the past, there has been a complex mix of weak linkages between transport supply and demand; some managed through various semi-regulated private sector structures and some managed wholly through a political process. The general assumption, although probably impractical, was that transport supply could be maintained through public funding at a quality level which roughly met the perceived demands of the population. The main focus of transport analysis was therefore on transport demand.

However, the demands of the population have increased to a level that cannot be accommodated in physical terms or matched by public funding, and this has been a key factor in creating pressure for transport policy changes. Within the new transport policies, it is recognised that strong policy linkages are needed between transport supply and demand, and that these should be based upon wider economic, social and environmental objectives.

Accessibility measures describe the links between transport supply and these wider policy areas. Considering them more explicitly within appraisal has the potential to define more clearly how transport policy objectives can be delivered through practical policies, programmes and projects.

In the past, qualitative accessibility concepts have been widely reported within appraisal, particularly where political aspects of evaluation have been important. In some situations simplified quantitative approaches have been adopted but the extent of this has been limited by concerns about the double counting of benefits. Although accessibility measures have been calculated as part of the demand estimation, they have rarely been presented as useful results in themselves. There are several reasons for this including:

- many of the benefits of transport investment are demonstrated through changes in travel patterns and demand;
- transport is a derived demand so the natural focus for analysis of transport needs has been transport demand rather than transport supply;

- there has been caution amongst policy-makers about emphasising the relationships between transport supply and demand. It was not until 1994 in the UK that some policy-makers formally recognised the relationship between road supply and traffic demand in the light of the results of a study by the Standing Committee on Trunk Road Assessment (SACTRA).

However, recent research has identified the role and application of robust quantitative approaches, which will allow accessibility methods to take a more central role within transport appraisal overcoming constraints with current methodologies (Scottish Executive, 2000). This paper discusses the need for accessibility analysis, and how practical quantitative analysis can be undertaken to support policy, programme and project appraisal.

The Need for Accessibility Analysis

The economic and social welfare for any individual is dependent upon the opportunities or choices available to them. The primary aim of transport is to enable people and businesses to gain access to jobs, shops, friends, family, and many other activities. The ease with which these activities can be accessed depends not just upon transport systems but upon patterns of land use. Accessibility measures seek to define the level of opportunity and choice taking account of both the existence of opportunities, and the transport options available to reach them.

Accessibility analysis is therefore needed to assist with many transport decisions by:

- ensuring consistency between transport policy and other public policy objectives including land use, health, education, and regional development;
- establishing the effects of changes in the transport system (i.e. including all modes, interchanges, cost, time, reliability, and quality) on people's access to opportunities such as employment, shopping, health services, social support networks, recreation, countryside etc.;
- defining how transport impacts are distributed across geographical areas, population groups, trip purposes and modes of travel including compatibility with equity objectives;
- assessing the impacts of new developments and projects;
- examining local access opportunities by walking and cycling including access to public transport;
- quantifying the value of additional travel options for each sector within society.

There is also a need for transport evaluation to be understandable. One impact of increasing complexity in transport appraisal has been for some policy-makers to undertake no analysis and make decisions on an intuitive basis. Given the importance of transport to economic, social and environmental aims this is a major concern. Decision-makers want to know what the impact of a transport programme or project will be on issues such as access to jobs, health, education, leisure, and retail opportunities before they commit significant public funding. Accessibility analysis can be as simple or as complex as required for each situation from a simple catchment assessment (i.e. *x* jobs within *y* minutes travel) to rigorous economic analysis.

The need for, and use of, accessibility analysis in remote areas has been established for some time. In these areas, travel demand has rarely been as important as accessibility in policy terms. The practical experience of appraisal of transport in these areas (Halden, 1994) can be applied more generally. Peripherality measures are examples of how quantitative accessibility analysis can be used to specify and evaluate how policies can best be delivered (Copus, 1999).

For most current evaluation, transport demand modelling is a direct input to the policy, programme and project appraisal. Figure 13.1 illustrates this approach.

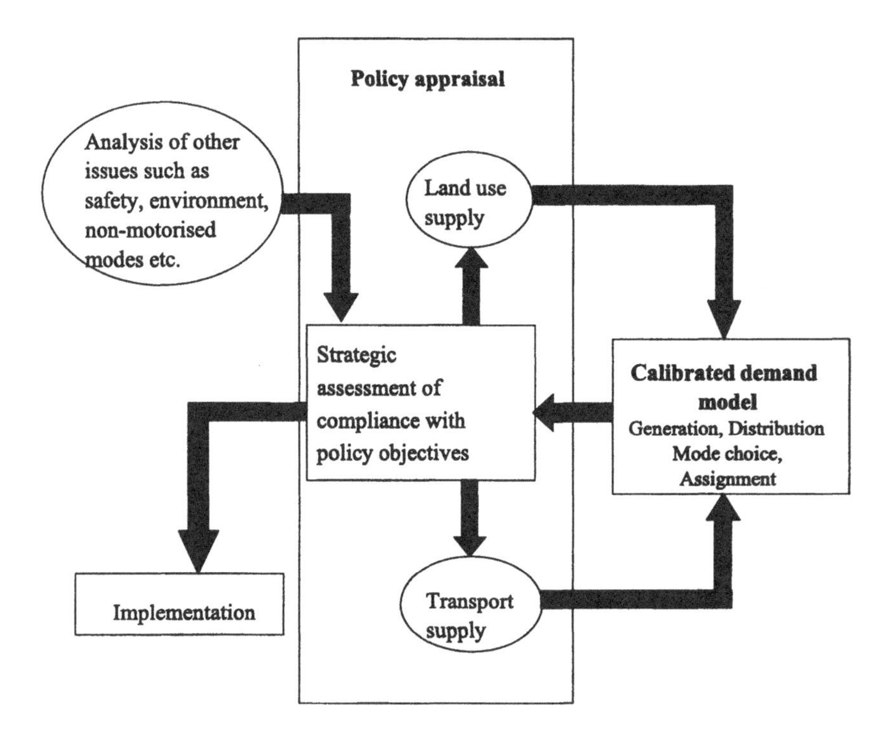

Figure 13.1 Policy, Programme and Project Appraisal

If demand is to be managed through a more integrated policy approach then demand modelling continues to be an important input to the analysis but it is viewed within the context of strategic accessibility objectives as shown in Figure 13.2.

This ensures that a consistent approach is taken to all modes including walking and cycling (where demand is not a major influence on accessibility). It also allows a much wider range of policy impacts to be studied since as many user groups as is appropriate can be investigated without increasing the complexity of the demand modelling.

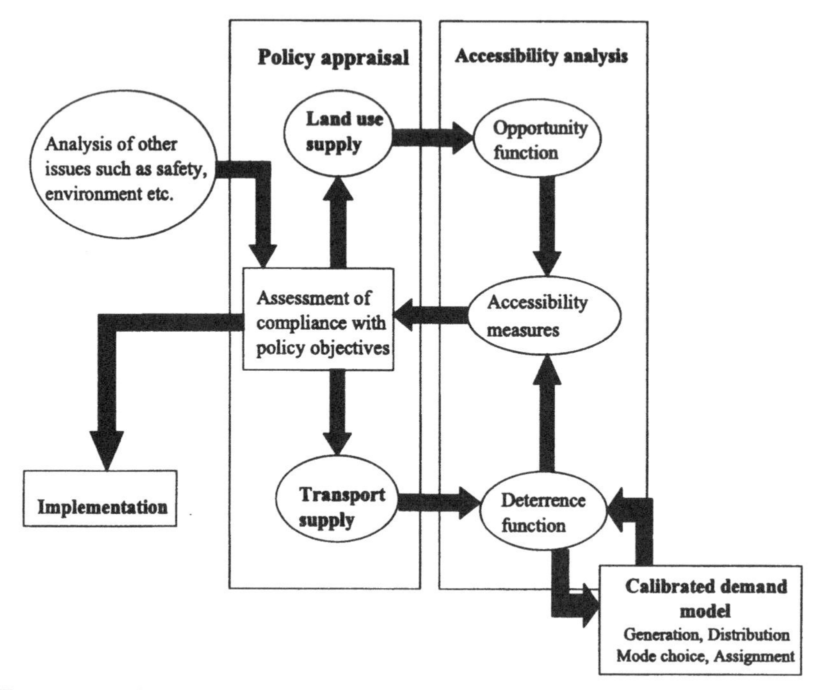

Figure 13.2 Strategic Accessibility Objectives

Components of Accessibility

All accessibility measures relate to a specific location, origin or destination, and include representation of defined opportunities and a separation element between these opportunities and the location. As noted above, the complexity of the accessibility analysis needs to be appropriate for the needs of the particular situation. The *opportunity* terms, *deterrence* functions and the sizes of the *zones* for considering accessibility therefore need to be defined at an appropriate level of detail. Before looking in more detail at the different types of measures, some comments are made on these three elements of accessibility analysis.

Opportunity Terms

The type of opportunities depend upon whether origins or destinations are being considered:

- origin accessibility considers the opportunities available to an individual or a business. The opportunity term is therefore usually based upon the land uses at alternative destinations;
- destination accessibility considers the catchments for a destination. The opportunity term is therefore usually based upon the land uses and type of person or traveller at alternative origins.

Land uses of interest include:

- employment, education and training – employment locations, schools, colleges, universities, training centres;
- health and social-health centres, hospitals, social security offices, job centres, post offices;
- shopping and leisure-shops/shopping centres, cinemas, theatres, sports centres, outdoor activity opportunities, centres for religious activity, pubs, clubs.

Types of person or traveller take account of:

- mobility – car ownership, disability;
- employment status – unemployed, economically active etc;
- age – retired, adult, children, etc.

The land use data and type of person can be obtained from national, regional and local statistics to the required level of accuracy.

Deterrence Functions

The deterrence function can be measured as time, travel cost, distance, or generalised cost/time. It aims to represent real behaviour and perception of travel. This must include the relative deterrent effect of different types of travel, and the costs associated with each, including issues such as the greater deterrent effect of time waiting for a vehicle when compared with the same time spent travelling in a vehicle.

It is usually helpful to look separately at the deterrence functions for car available and non-car available trips. Many trips will involve a combination of several modes, and for non-car available trips the car options are excluded from the calculation. For example a car available trip to a city centre from a rural area may involve a car element to a park and ride site, a bus element from the edge of the city to the centre and a walk element from the bus terminus to the destination. The non-car available alternative would consider only the public transport, walking and cycling options to reach the city centre.

However each trip has other characteristics which can make generalisation for the purpose of analysis difficult. The reason for not making a walking or public transport trip may be the need to carry goods, the need to take other people, the weather, the perceived quality of the route including personal security and safety considerations, or simply a lack of knowledge of available options. All these factors can be affected by transport policy decisions, so it is desirable if appraisal can take account of them in a meaningful way.

To ensure a robust approach, calibration against observed behaviour should provide a firm foundation on which to build. The accuracy of the calibration is heavily dependent upon the quantity and quality of the travel survey data, and this can be expensive to collect. However, data availability on travel patterns is improving in the UK as part of many local transport plans so this problem may be less significant in the future.

One further aspect demands comment. Travel patterns are not static, so observations of travel behaviour should ideally take account of trends in trip making rather than simply observed demand. There is no reason in principle why deterrence functions and accessibility indices should not be able to incorporate these more dynamic relationships. Nevertheless, such techniques have rarely been adopted in practice to date although they have been shown through research to have considerable potential (e.g. Levinson, 1995). In the meantime, dynamic accessibility analysis must therefore be considered as a future aspiration rather than a practical prospect.

Zones

The extent of the zoning system and the level of detail will depend upon the policy issues being examined and how much effort can be afforded on the analysis. Strategic transport improvements will require a wide geographical coverage, but a fairly coarse zoning system may be adequate, whereas a local issue such as the accessibility of a school will require very detailed local representation.

Reliable data are usually easier to obtain for coarser zones and in practice accessibility analysis will often be able to adopt zoning systems defined within established transport demand models which have generally been designed to take account of the geography of transport networks.

Methodologies

Over the last 20 years there have been comprehensive reviews of accessibility theory (Jones, 1981; Simmonds, 1998) but only limited steps have been taken towards practical application of the techniques. The recent Scottish review (Scottish Executive, 2000) identifies that for the purposes of the practical application of these measures, there are three generic but overlapping types of indicator, which can be described as:

- simple indicators – with these, the representation of transport and/or opportunity within the accessibility equation is simplified by defining thresholds (e.g. number of relevant opportunities within a given travel cost, time, etc.; measures of the travel cost, time, etc. required to reach a given number of opportunities; shopping or employment opportunities with more than a defined floor space or number of jobs etc.);

- opportunity measures – these sum up all the available opportunities and weight them by a measure of deterrence based upon how easily the opportunities can be reached;
- value measures – these seek to define the attractiveness of the available opportunities to represent their value as a transport choice;

Simple Measures

The time and distance thresholds adopted within these measures will depend upon the policy issue. Thresholds can be applied to either the opportunities (i.e. shopping centres with more than 5,000 square metres floor space) or the deterrence functions (e.g. jobs within 60 minutes travel time) or both.

The accessibility measure for a location (i) is calculated as the sum of the opportunities available at alternative locations (j) within the defined threshold.

$$A_i = \sum O_j \delta_{ij}$$

Where $\delta = 1$ if the opportunity is within the generalised time or cost threshold, and $\delta = 0$ otherwise.

Generalised time is often used, since greater inaccuracy is likely converting time to generalised costs than in converting fares and other costs to time using appropriate values of time for the user group. However generalised cost can be used effectively and this will often be more appropriate where economic analysis is being considered.

The choice of thresholds must accurately reflect some aspect of travel behaviour for these measures to be useful. For local access by walking, a five minute walk equates to about 400 metres, a 10 minute walk to about 800 metres and a 20 minute walk to about 1600 metres. These can be taken to represent thresholds for a short walk, a normal walk and a maximum walk respectively. Beyond the 1600 metres threshold, very few trips are made by walking (Ecotec, 1993). Recent guidance in England (DETR, 2000) also identifies 250 metres as a walking threshold for access to public transport.

For more strategic destinations, thresholds based upon observed behaviour are harder to define, but the use of a range of values such as 15 minutes, 30 minutes and 60 minutes can give useful information. However care needs to be exercised. A major opportunity which is 31 minutes from an origin in a base situation may fall within a threshold within a design situation giving a misleading impression about the real impacts of a transport change. Decisions related to thresholds must ultimately be defined using behaviour specific to local characteristics.

Opportunity Measures

The Opportunity measure for a location (i) is calculated as the sum of the opportunities available at locations (j) factored by a deterrence function based upon the travel time between i. and j.

$$A_i = \sum O_j \exp(-\lambda t_{ij})$$

Where exp(-λt) is the deterrence function, and O_j is the opportunity available within zone j.

The accessibility of zone i. is the total opportunity, with the units being the number of jobs, retail opportunities, etc. In order to achieve the correct sensitivity of the indices to travel time, the deterrence function is calibrated using the λ factor. In many situations the deterrence function will have been calibrated against observed travel patterns to achieve accurate trip distribution.

However, accessibility analysis can still be useful without local calibration of the deterrence function, since default values of λ by trip purpose can be used to give meaningful results.

Value Measures

The Value measure for a location uses the same input data as for the opportunity measures but the equation has been transformed so that it represents the value of the opportunities in (generalised) time, or cost. The relatively simple form of the Utility index used in recent Scottish studies was:

$$A_i = \frac{1}{\lambda} \ln \frac{\sum [\exp(\lambda t_{ij}) O_j]}{\sum O_j}$$

It is also worth noting at this stage the potential for utility indices to be used in economic analysis combining the consideration of benefits for motorized and un-motorised travel. Simmonds (1998) notes that 'properly constructed composite utilities represent the best measure that transport economics has so far devised to measure the overall ease or difficulty of travel from a particular place, by particular groups of people, for a particular purpose, and that the formula that is obtained from measuring changes in accessibility in this way is identical with the one used to measure changes in user benefit'. These issues are discussed in more depth below.

Types of Measure

Table 13.1 shows how many of the commonly used indices are categorised under each of these three main types identified above.

Table 13.1 Types of accessibility measure

Index	Description and Uses
Simple Measures	
Catchment/Contour indices	These count the number of people, jobs, shops etc within a threshold travel cost (distance, time etc.) from a defined location. They are used for a wide variety of planning purposes for both land use and transport infrastructure and are often used by developers to consider the potential commercial viability of a potential development location.
Access to public transport	Rather than looking at transport network accessibility to destinations, these measure walking access time to the public transport services themselves. Walking time or distance thresholds to the public transport services are set and summed across all the available services. The quality of public transport being accessed is categorised on a scale which takes account of service frequency, type of service (i.e. rail/bus/light rail etc.) and service reliability. Although of limited scope, the simplicity of this approach has proved attractive and the calculation and mapping procedures have been automated and marketed by various organisations.
Peripherality indices/Rural accessibility	These identify thresholds in terms of cost, distance, time etc from defined types of opportunity. These are usually calculated from major centres of population such as towns or cities or public services such as hospitals, but have also been used to study accessibility to transport networks including the European Community Trans European Networks.
Time space geographic measures	These measures simplify travel behaviour and choice in terms of the opportunities available within a limited travel time budget. The threshold is therefore the travel time available for a particular individual or group. These are widely used in logistics planning for freight but are equally applicable to people accessibility issues.
Opportunity Measures	
Hansen indices	The simple measures above are all special forms of Hansen indices incorporating thresholds to simplify data or analysis requirements. Hansen indices have had wide application within research and are used within transport models to estimate trip distribution.
Shimbel measures	These are a specific case of the Hansen indices in which all specified opportunities are assumed to have the same weighting. The measure is simply the sum of the cost (time etc.) to each of the opportunities.
'Economic potential' measures	Where the opportunities being considered in the Hansen index are regional incomes, and the deterrence function is measured in distance, then the accessibility index is sometimes described (Keeble, 1982) as the economic potential of a location.
Value Measures	
Utility based measures	These measure the value to an individual or group of the choices available to them. The main difference from the opportunity measures is that the indices are less sensitive to large changes in number of opportunities so are relatively more sensitive to transport changes. The normal units of measurement are generalised cost or time and these measures are widely used within transport models.

Each of the above measures can be expressed in many different ways. The approach chosen will reflect the needs of the particular situation. At its simplest level qualitative descriptions can be used to define the accessibility of a location.

Terms such as town centre, rural area, remote area, and 'accessible location' can be used as simple qualitative accessibility measures describing locations in terms of the population, availability of local opportunities, and sometimes transport supply. However analysis to support practical decision-making will usually benefit from a more rigorous approach so qualitative (i.e. good, average, poor) or quantitative indicators can be easily used within multi-criteria framework analysis.

Uses

Accessibility analysis has been widely used in the past, but increased rigour in the approaches, combined with further development of the techniques could allow it to play a more central role. A particular strength of accessibility analysis is that it allows transport to be viewed from the perspective of users choices, in addition to providing a framework for transport policy, programme and project evaluation.

Success in developing integrated transport depends upon *building support from potential users* in addition to *effective planning of transport networks and systems*. It is therefore relevant to look at the role of accessibility analysis for both these purposes.

Understanding and Influencing Travel Behaviour

Transport planners often view transport planning in terms of coping with or managing demand, since transport is their primary focus and their main area of responsibility. However, other people including travellers view the quality of transport in terms of the opportunities and choices available to them. This involves combining transport and land use within the analysis using accessibility measures to describe the opportunities and choices.

One of the greatest uncertainties in transport planning is the estimation of travellers' responses to changes in transport systems. By seeking to understand the factors affecting these choices in terms of accessibility transport planners can gain insights into the factors which affect these choices.

It has been proposed (Halden, 1996) that there are five underlying factors affecting travel choices: the quality and type of opportunity at the destination, and the travel time, cost, reliability and image. For any project, changes in travel behaviour as a result of investment should be explainable in terms of the changes in one or more of these parameters. Using this approach there may be significant potential for accessibility measures to identify reliable and transferable trends in travel behaviour not identifiable with more disaggregate variables. Also, any process to manage changes in travel behaviour must influence peoples' perception of their travel choices (Kelly, 1955). Accessibility measures offer a convenient tool to analyse the sensitivity of travel choices to changes in each parameter.

For example 'network effects' may have been underestimated in transport appraisal in the past (SACTRA, 1999) largely due to the poor understanding of the importance of image and reliability issues. Whilst it is likely to continue to be difficult to establish reliable and transferable trends between image and reliability and network effects, relationships between accessibility and networks may be more manageable.

The need for better understanding of accessibility and travel behaviour is becoming increasingly important with the continued development of electronic telecommunications. The general approach to accessibility analysis covers all choices including options that avoid the need to travel, such as video conferencing. Much better understanding of the impacts of new ways of working could be achieved by researching how people perceive electronic access opportunities relative to accessibility through travel.

However increasing electronic communication is just one uncertain factor within the fast changing and complex technical and administrative structures within the EC. Many evaluation techniques do not lend themselves to the evaluation of uncertainty, but accessibility indices used in conjunction with risk assessment techniques can be used to manage uncertainty for current and future scenario. Where accessibility is sensitive to a parameter within the analysis, action can be taken to manage this uncertainty in the implementation of the change.

Political decisions about transport investment are often more influenced by public opinion than by technical appraisals, particularly at a local level. Therefore the ability of transport policy appraisal to influence public opinion is very important. Accessibility analysis can have major presentational advantages by describing the impacts of transport investment in terms that people can easily understand such as changes in the numbers of jobs accessible from a given location.

Good accessibility as a transport objective carries very broad support from every strand of opinion within society. Accessibility measures can be presented to ensure transparency about the impacts of transport proposals, so are particularly helpful in building consensus about how to improve transport. If people understand how they can benefit from investment decisions they will be more prepared to change their behaviour to take advantage of the new opportunities.

Policy, Programme and Project Evaluation

There are many levels of decision-making about transport policy, programme and project evaluation. Accessibility analysis can be used at each of these levels with the detail and accuracy of the analysis appropriate for the needs of the particular decisions. For example:

- at an inter-regional level, accessibility analysis could be used to define overall performance levels for transport and targets for improvement, including for strategic road and rail networks, in addition to checking that local and regional strategies are consistent with economic development and social objectives;
- at a regional level, accessibility analysis could be used to check the impacts of joint plans or specific schemes (such as congestion charging) geographically or on types of traveller;
- at a local level, accessibility analysis can demonstrate the extent to which transport proposals are consistent with strategic plans and support economic, social and environmental policy objectives.

At all levels the analysis can take an integrated approach to the evaluation of transport in relation to land use, economic development, social inclusion/equity, and other policies. Provided information is available at an appropriate spatial scale on houses, jobs, population sector etc. then the transport issues can be combined with these data within the accessibility analysis.

As discussed above the results of accessibility analysis can be presented in terms of generalised cost or time allowing the economic value of a particular policy, programme or project option to be assessed. This has many advantages as follows:

- comprehensive composite accessibility analysis provides a complete measure of the economic value of a particular pattern of land use and transport supply. Travel demand models are always less than fully comprehensive so once these accessibility methods become established they may provide a more accurate assessment of the economic value of transport improvements;
- accessibility analysis considers the spatial impact of change so that economic development issues are explicitly considered;
- accessibility analysis provides robust economic analysis of transport projects within which motorised and non-motorised modes are treated consistently;
- by looking at the value of transport *supply* to people, rather than demand for travel by people, the analysis includes important aspects of the value of transport systems which would be excluded from a demand analysis. These 'transport option' values include: the value of knowing that a bus will be available if a car breaks down, the ability to access new clients or suppliers if needed, and the value of choice in terms of destination or time of travel;
- accessibility analysis helps to inform judgements on the indirect consequences of transport changes on issues such as land uses and car ownership.

The main problem with the wider application of accessibility concepts for economic analysis is that to date the techniques have rarely been used outside research. Practical application of accessibility analysis has generally been confined to simple quantified or qualitative techniques. However some useful applications have been undertaken on economic development studies (Vickerman, 1995).

The recent Scottish research considered four case studies in Scotland. Although economic analysis was specifically excluded from the scope of this work, integration with other policies such as social inclusion and land use was a key element. The work identified how the policy, programme and project issues in relation to each study area could be evaluated. Key findings include:

- accessibility analysis focuses on people rather than modes making it particularly suitable for appraisal within integrated transport policies. The impacts on any group of people can be quantified and comparisons between different groups made;
- minority groups can easily be considered in the analysis so that answers to a very wide range of policy questions can easily be obtained (e.g. accessibility to employment for non-car available unemployed people);

- it can be helpful to evaluate impacts on equity by using ratios of accessibility for different mobility groups including car available, non-car available and mobility impaired people;
- accessibility analysis techniques need to be progressively implemented more widely within practical land use and transport appraisal. This will allow issues such as data collection and presentation of results to be streamlined for policy, programme and project evaluation. Also standardised approaches can be developed to allow robust representation of travel behaviour for different types of trip without the need for detailed demand modelling;
- quantitative analysis methods can be tailored to the particular policy, programme or project requirements based on the three main types of index. *Simple measures* are easy to understand and are most useful for local walking and cycling accessibility including assessing access to public transport services. *Opportunity measures* also have the benefit of being easy to understand since, like the simple measures, they are expressed in terms of number of jobs, number of people etc. They have many potential uses including: the comparison of accessibility changes for different population groups, the identification of catchments for key trip destinations, and the comparison of accessibility for car available and non-car available trips. ***Value indices*** can be more difficult to interpret but provide a direct measure of the value of transport systems with the potential for future development in economic appraisal;
- the choice of zoning system for the accessibility analysis must be appropriate for the policy or scheme being evaluated. Robust data are easier to obtain if a coarse zoning system is adopted, so transport networks can be modelled at a strategic level without excessive difficulty.

Practical Application and Presentation

As noted above, accessibility measures are already used extensively in transport analysis but they have not yet been used to their full potential. One of the main reasons for this is that despite the policy, presentational and practical advantages of the approach, care is needed to ensure robust and comparable results.

For practical presentation of results, the Scottish research (Scottish Executive, 2000) suggests five main categories:

- transport 'system' accessibility to opportunities such as jobs, education, etc;
- ratios comparing accessibility for different groups;
- accessibility by walking and cycling to local facilities;
- accessibility by walking and cycling to public transport services;
- accessibility for freight.

Table 13.2 summarises the Main Applications for each Analysis Category.

Table 13.2 Application of Analysis Categories

Category	Application
Transport 'system' accessibility	Origin accessibility – Analysis of the employment, shopping, leisure etc. opportunities available to people from each mobility group. Destination accessibility – Defining the people and business within the catchment of a defined location.
Ratios comparing accessibility for different groups	Origin accessibility – To compare the choices available for different mobility groups. Destination accessibility – To define the transport characteristics of a location for development planning e.g. out of town, city centre.
Accessibility by walking and cycling to local facilities	Origin accessibility – Looking at the opportunities available to people in each area without the use of motorised transport. Destination accessibility – Identifying the catchments of local facilities such as health centres, town halls, sports centres, shops etc.
Accessibility by walking and cycling to public transport services	Origin accessibility – Identifying how well public transport serves peoples' needs. Destination accessibility – Most suited to the planning of bus routes and development planning in dense urban areas.
Accessibility for freight	Origin accessibility – Companies planning their logistics operations. Destination accessibility – Classification of type of destination in terms of function, catchment, and type of freight.

Data management can prove to be a challenge for large applications. Various techniques are in use to overcome these challenges. Modern computer databases can easily be adapted to accessibility analysis requirements. These allow significant changes to be easily identified and reported graphically if required. Geographical Information Systems (GIS) have also been extensively used for analysis and presentation. They have the advantage of easily being able to assimilate point data such as the location of a hospital with zonal data. Transport systems need to be represented in terms of the ease of travelling between zones but much of the land use data available is most easily available within GIS at a point level.

Figure 13.3 shows accessibility from different parts of Edinburgh to the new Royal Infirmary which is currently under construction. The darker shades show good accessibility and, by testing alternative transport strategies such as the introduction of new bus services, the geographical distribution of impacts is observed directly.

Such analysis is currently being undertaken for a wide range of existing and proposed land uses throughout the city to inform the development of integrated land use and transport plans. In addition to considering the opportunities for the population as a whole for access for work, shopping, education, healthcare etc. the analysis is considering the impacts of proposed changes on different sections of the population including unemployed people, deprived households, ethnic groups etc.

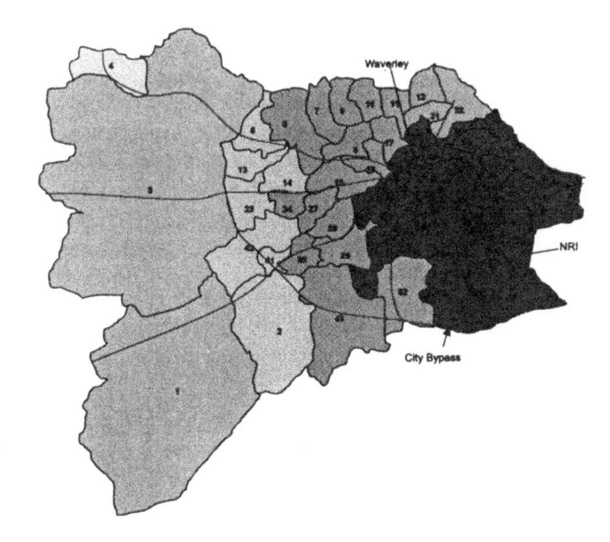

Figure 13.3 Non-car Accessibility to the New Royal Infirmary in Edinburgh

Through studies such as this, experience of accessibility analysis is growing and optimum approaches are being identified to simplify data collection and analysis. There continues to be a need for more practical applications to further demonstrate good practice and to develop improved methods for presenting results.

Conclusions

There are many strong reasons why policy, programme and project appraisal for integrated transport should be able to benefit from accessibility analysis, since accessibility measures describe the opportunities and choices available for people and businesses. Although current practice relies mainly upon qualitative approaches, increasing use of quantitative methods will be required for the full potential benefits of accessibility analysis to be realised.

Although there is a large range of accessibility measurement techniques available, there are three generic types:

- simple indicators;
- opportunity measures;
- value measures.

All three types of indicator have a role to play in policy and scheme appraisal, since different decisions require information to be presented in different ways.

References

Copus, A. (1999), *A New Peripherality Index for the NUTS III Regions of the European Union*, Report for DGXVI.
Dasgupta, M. and Sharman, K. (1994), 'Transport and Urban Change: Commuting Trends in 27 British Cities and Towns', *Project Report PR/TR/023/94*, Transport Research Laboratory, Crowthorne.
Department of the Environment and Department of Transport (1995), *PPG 13 – A Guide to Better Practice*, HMSO.
Ecotec (1993), *Reducing Transport Emissions Through Planning*, Department of the Environment and Department of Transport, HMSO, London.
Halden, D., Consultancy (2000), 'Review of Accessibility Analysis Techniques and their Application', *Scottish Executive*, Edinburgh.
Halden, D. and Sharman, K. (1994), 'Transport and Development around Inverness', TRL Project Report TR/SC/07/94.
Halden, D., Emmerson, P. and Gordon, A. (1995), 'Skye Bridge Socio-Economic Impact Study – Development of Strategic Transport Modelling and User Cost Benefit Analysis', TRL Project Report TR/SC/29/95.
Jones, S.R. (1981), 'Accessibility Measures: A Literature Review', *Transport and Road Research Laboratory*, Report 967.
Keeble, D. , Owens, P.L. and Thomson, C. (1982), 'Regional Accessibility and Economic Potential in the European Community', *Regional Studies*, No. 16.
Kelly, G. (1955), *The Psychology of Personal Constructs*, Norton, New York (reprinted 1991, Routledge, London).
Kerrigan, M. (1992), 'Measuring Accessibility – A Public Transport Accessibility Index', PTRC.
Levinson, D. (1995), 'An Evolutionary Transportation Planning Model – Structure and Application', *Transport Research Board*, Washington DC.
LPAC London Planning Advisory Committee (1994), *Advice on Strategic Planning Guidance for London*.
McKinnon, A.C. (1989), *Physical Distribution Systems*, Routledge, London.
Simmonds, D. (1998), 'Accessibility as a Criterion for Project and Policy Appraisal', unpublished Report for the Department of Environment, Transport and the Regions, Consultancy, University of Leeds, MVA, Oxford Brookes University.
Simmonds, D. (2000), 'Environmental Resources Management Methodology for Multi-Modal Studies', Department of the Envirnoment, Transport and the Regions, Consultancy, John Bates Services, MVA, University of Leeds,
(2000), *Environmental Resources Management Methodology for Multi-Modal Studies*, Department of the Environment, Transport and the Regions.
SACTRA (1994), *Trunk Roads and the Generation of Traffic*, The Standing Advisory Committee on Trunk Road Assessment, Department of Transport.
SACTRA (1999), *Transport and the Economy*, The Standing Advisory Committee on Trunk Road Assessment, Department of the Environment Transport and the Regions.
Vickerman, R. (1995), 'Location, Accessibility and Regional Development, The Appraisal of Trans-European Networks', *Transport Policy*, London.

Chapter 14

Strategic Environmental Assessment and its Relationship to Transportation Projects

Paul Tomlinson and Chris Fry

Strategic Environmental Assessment

Strategic Environmental Assessment (SEA) can be defined as the formalised, systematic and comprehensive process of evaluating the environmental impacts of a strategic action and its alternatives, including the preparation of written reports on the findings of that evaluation, and the use of the findings in publicly accountable decision making. The term strategic does not relate to a specific geographic scale nor does it imply a degree of superficiality in the assessment activities. The 'strategic action' will normally relate to:

- *Policies:* Guidance drawn up by Government with a defined strategy.
- *Plans:* A set of co-ordinated and timed actions for the implementation of a policy in a particular sector or area.
- *Programmes:* A set of actions in a particular sector or area.

While these terms are clear as abstract concepts, they take on different meanings in different countries since their precise application is dependent upon individual political and institutional frameworks (Sadler and Verheem, 1996). Hence the term SEA is often applied to the evaluation of studies establishing alternative alignments within given corridors (corridor studies) as well as to plans and programmes.

Although considerable interest is now being shown in SEA, not least because of the European Commission Directive (EC, 2001, p.42), SEA is already being used in the United States where it is often termed a Programmatic or Regional EIS (Webb and Sigal, 1992). In New Zealand, the Resource Management Act 1991 provides for the provision of a statement of the environmental, social and monetary costs of policies and plans (Wood, 1992). While in Hong Kong a 1988 Government directive required plans to be subject to environmental assessment and there have been at least 10 major transportation or land use SEAs undertaken (Au, 1999). The People's Republic of China and Tanzania are also applying the principles of SEA. This focus upon SEA is leading to the development of practical methods, which is also revealing real constraints and issues that need to be resolved.

The SEA Directive

The concept of a SEA is not new. In fact early drafts of the first EIA Directive included an assessment of policies, plans and programmes as well as projects. For various reasons the SEA component was omitted, but the concept of SEA was not ignored. In March 1999 a draft Directive was published on the 'Assessment of the Effects of Certain Plans and Programmes on the Environment' (COM, 1999, p.73) – (the SEA Directive). With the Directive being published in June 2001, Member States have three years for implementation.

The Directive introduces a requirement for formal consideration of environmental issues in the production and adoption of plans and programmes which establish a framework for future development consents. It requires that in the preparation of such plans and programmes a study is made of the probable significant environmental effects of implementing the plan or programme or alternative proposals. An Environment Report and Non-Technical Summary is prepared and made available. Consultations with relevant bodies and the public are to be carried out and the results of the study and consultation are taken into account prior to the formal adoption of such plans or programmes.

These requirements apply to plans and programmes which are prepared and/or adopted by 'competent authorities', i.e. central or local government bodies. In the case of transportation planning in the UK, it is envisaged that the SEA Directive would apply to national transport investment programmes, regional transportation strategies as well as local transport plans or strategies. Private sector transport operator plans may be included in so far as they relate to rail, port and airport plans that are a requirement of legislation. Modifications to existing plans and programmes would also fall under the scope of the Directive where they are likely to have significant negative environmental effects.

Under Article 5 of the SEA Directive, the likely significant effects of implementing the plan or programme are to be identified, described and evaluated, although it is through Annexes I and II that the characteristics of the effects and the area likely to be affected are detailed. The Environment Report is to take account of the level of detail in the plan or programme, its stage in the decision-making process and the extent to which certain matters are more appropriately assessed at different levels in order to avoid duplication. This point will form the basis for further exploration later in the paper.

The information to be provided in an Environment Report is defined in an Annex to the Directive as being:

- an outline of the contents, main objectives of the plan or programme and relationship with other relevant plans and programmes;
- the relevant aspects of the current state of the environment and the likely evolution thereof without implementing the plan or programme;
- the environmental characteristics of areas likely to be significantly affected;
- any existing environmental problems relevant to the plan or programme;
- the environmental protection objectives which are relevant to the plan or programme and the manner in which they have been taken into account;
- the likely significant effects on the environment, including human health;

- the measures envisaged to prevent, reduce and as fully as possible offset any significant adverse effects;
- an outline of the reasons for selecting the alternatives dealt with, and a description of how the assessment was undertaken including any difficulties;
- a description of monitoring measures and
- a non-technical summary of the information provided under the above headings.

Article 7 deals with consultations on trans-boundary effects, and under Article 8 the competent authority should take into consideration the results of all consultations. Article 9 then stipulates that the competent authorities shall make available to the public and environmental authorities consulted a copy of the adopted plan or programme and a statement of how the results of the consultations have been taken into account. These arrangements are for Member States to define.

An interesting inclusion is a requirement under Article 9c that information is to be provided on the measures to be used to monitor implementation of the plan or programme. This proposal is to identify unforeseen adverse effects and to take remedial action as necessary.

Article 12 contains a provision via clause 2 that requires Member States to ensure that appropriate measures are taken to guarantee that the Environment Reports meet the minimum requirements and to prevent decisions being taken where they are not reached. As the plan proponents are also likely to be those responsible for commissioning or undertaking the SEA, there could well be tension between the plan making activities and a robust SEA. Perhaps this may lead to the formulation of SEA review organisations independent from the plan making organisations.

One clause sadly missing from the Directive is any explicit linkage between SEA and project EIA. At the very least one might have expected a clause that requires project EIA to be in broad conformity with the relevant spatial or sectoral SEA.

Suitability of Transportation Planning for SEA

It is increasingly recognised that decision-makers need to be equipped with information that adequately informs them of the consequences of addressing the demand for transport and the extent to which the objectives of sustainable mobility can be met.

New Policy Environment

The UK Roads Review *A New Deal for Trunk Roads in England* (DETR, 1998a) and the *New Approach to Appraisal* (DETR, 1998b) provided the first insight to the changing balance of interests in highway planning. In this new approach land use planning, environmental assessment and public involvement are no longer to be seen as after the fact add-ons, but are to be fully integrated and equal activities to that of transportation planning.

The UK Government has also introduced new procedures and a variety of new guidance on assessment methodologies for assembling information for decision-makers. Through the New Approach to Appraisal (NATA), the concept of a one page tabular summary of the main environmental, economic and social consequences of a road investment proposal – the Appraisal Summary Table – was launched (for example see DETR, 1998c).

Guidance on the Methodology for Multi-Modal Studies

The appraisal process set out in the *Guidance on the Methodology for Multi-Modal Studies* (GOMMMS) (DETR, 2000) builds upon that of the *New Approach to Appraisal*. While GOMMMS does not seek to meet the requirements of the SEA Directive, nor address the links with project delivery, it does mark yet another high-point for transportation planning. GOMMMS results in the following changes:

- *a Multi-Measure and Modal Focus:* It provides a methodology for a change from a road building mentality to one where multiple transportation measures involving multiple modes and non-infrastructure solutions need to be brought to bear on the transport problem;
- *clear exposition of objectives:* It recognises that highway design and transport planning objectives need to be set within a wider set of social, economic, environmental, accessibility and integration objectives;
- *clear definition of the transport problem:* It calls for a thorough understanding of the transport problem to be addressed recognising that the problem is to be set in relation to all of the objectives rather than simply in terms of traffic;
- *integrated appraisal:* The degree to which objectives are achieved is to be summarised along side the extent to which problems are ameliorated and the implications for distribution and equity, affordability and financial sustainability as well as practicality and acceptability.

Preparing for Strategic Environmental Assessment

Alongside the arrival of Multi-Modal Studies and GOMMMS, the Highways Agency (an agency of DETR[1]) embarked upon the preparation of a Guidance Manual on Strategic Environmental Assessment for Multi-Modal Studies which was led by the UK Transport Research Laboratory (TRL Limited).

In developing the guidance it was important to recognise the practical issues associated with Multi-Modal Studies which may extend over large geographic areas, as well as the preliminary level of detail to which the transportation measures are specified. Consequently, the SEA methodologies needed to be fundamentally different from those used in EIA. Essentially, the methodologies need to allow determination of the extent to which environmental objectives are achieved and also to ensure that data assembly requirements are realistic in the context of the uncertainties and assumptions associated with the transportation measures and plans being evaluated. An idealised view of the process is presented in Figure 14.1.

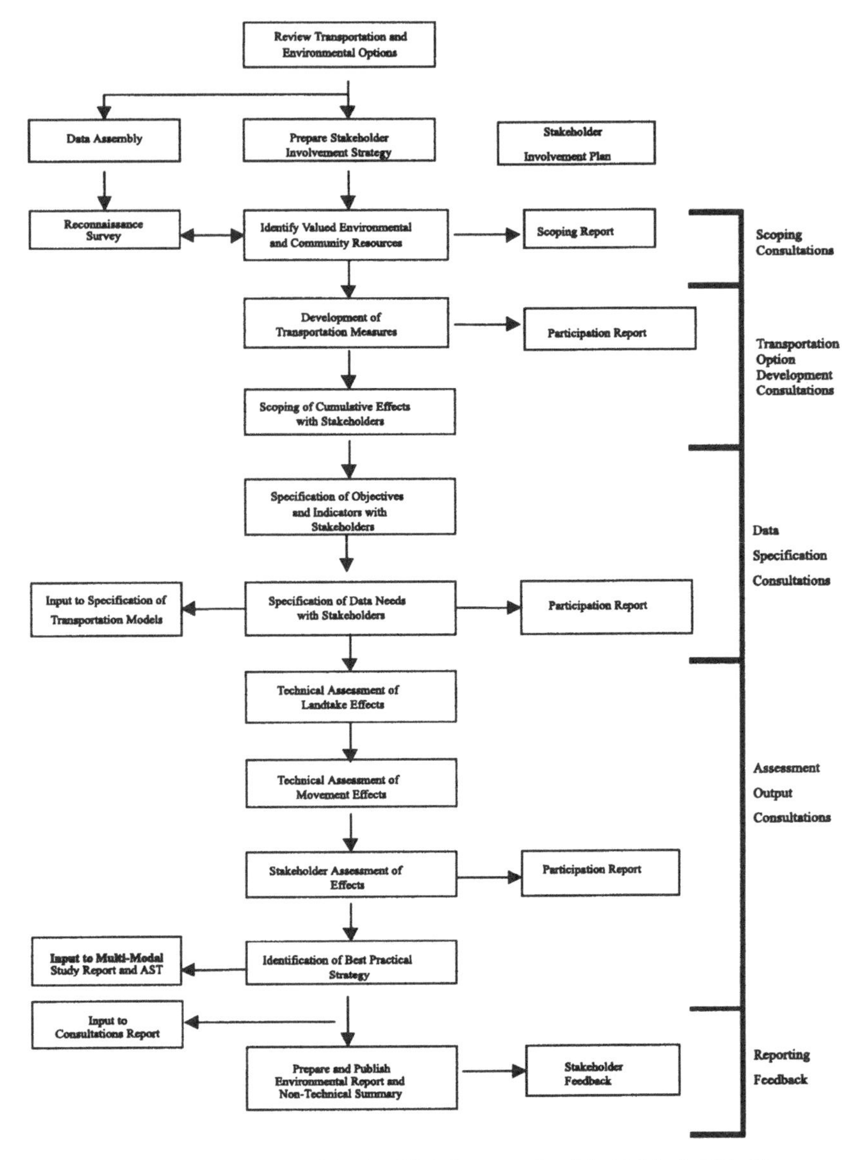

Figure 14.1 Idealised Activities in SEA Process for Multi-Modal Studies

Common SEA-EIA Methodologies

The structure of a SEA process depends on the planning procedure to which the SEA is linked. However the key steps would generally comprise:

- *Screening:* A process to determine whether SEA is required;
- *Scoping:* Determination of the issues that need to be examined, at what depth and in what way;
- *Establish environmental objectives:* As an objectives-led assessment, the objectives and their indicators need to be established to reflect the characteristics of each SEA;
- *Define measures to be assessed:* Unlike project EIA, the measures or actions being considered within SEA normally emerge part way through the process. For example it will not necessarily be known whether a new motorway, railway or a traffic management regime is to be assessed until an appreciation of the problems is completed;
- *Define potential effects:* Environmental objectives, awareness of valued environmental and community resources are used to identify potential areas of conflict with the transportation measures;
- *Establish environmental indicators:* Define indicators on which the environmental effects of the transportation measures can be assessed reflecting the agreed environmental objectives;
- *Establish future baseline:* Forecast future conditions without the potential transportation measures;
- *Impact identification and forecasting:* Determine the environmental change that results from the actions being examined;
- *Cumulative effects:* Identify and assess cumulative effects upon valued environmental and community resources;
- *Mitigation and enhancement:* Consider opportunities to provide mitigation or enhancement of transportation measures individually and collectively;
- *Assessment:* Assess the significance of the environmental impacts from each transportation measure and collectively;
- *Implementation and monitoring:* Identify mechanisms for addressing adverse effects and monitoring requirements;
- *Reporting:* Provide an overall assessment for the decision-making process and reporting to external audiences;
- *Stakeholder involvement:* Provide for input and review roles throughout the process.

While the broad assessment tasks of SEA are fundamentally the same as in project EIA, some significant differences exist. For example, SEA has a greater scope than EIA in that the geographical scale of a SEA tends to be greater than an EIA. Also the proposed action generally contains a number of different elements rather than a single project. Being at a plan or policy level, a larger range of alternatives is considered. The strategic scale also necessitates that a different range of environmental effects is addressed.

SEA is an objectives-led process in which environmental objectives provide the framework for testing the performance of the action. Consequently, there is less emphasis upon impacts unlike project EIA. A further difference is that as the time interval between planning, approving and implementation is much longer, the assessment process is subject to greater uncertainty. Also, the level of detail is generally less than that needed for project assessment.

Although the generalised process of SEA and EIA are essentially similar, SEA will always be fundamentally different as SEA requires greater simplicity, flexibility, adaptability, incorporation of value-judgements in order to address the higher levels of uncertainty inherent in the exercise. Consequently, SEA needs its own approach designed to meet the rationale of individual sector plan/programme activities.

Differences between SEA and EIA

Experience with EIA has demonstrated that its use in an individual transport project can be too late to tackle long-term, cumulative, global or policy issues, such as the effects of traffic growth and atmospheric emissions or changes in land use. Concern over the effectiveness and efficiency of project EIAs has often highlighted that project decision-making is constrained by decisions taken at higher levels where too little consideration is taken of environmental effects. Equally, where policy choices exist they cannot be adequately addressed at the project level. This also compounds the poor consideration of alternatives when a decision has been taken that a road represents the solution with little consideration of the other measures that may be complementary let alone a viable alternative.

Consideration of cumulative effects of the different elements of a transport infrastructure programme is difficult and indirect impacts are often overlooked. In particular, the interaction of transport and land use planning can be difficult to assess. Also as mitigation measures are being limited to the particular project being examined, so cumulative impacts across the network are often overlooked.

SEA is seen to address some of the weaknesses of project EIA with further support for SEA coming from the desire to show whether policies, plans and programmes aid sustainable development. The objectives for SEA, therefore, are to ensure the full consideration of alternative policy options, including the 'do-nothing' option, at an early time when greater flexibility exists. In considering alternatives in a more explicit manner than project EIA, increased emphasis is given to cumulative, indirect or secondary impacts of different multiple activities. Through early consideration of the environmental impacts, this allows adverse effects to be avoided or prevented.

SEA should also ensure that the environmental impacts of policies, that do not have an overt environmental dimension, are assessed, while at the same time minimising the prospect of needless reassessment of issues at a project level, thus saving time and money. SEA ensures that the environmental principles such as sustainability and the precautionary principle are integrated into the development, appraisal and selection of policy options. Thus it provides a proper place for environmental considerations in decision-making (Therivel, *et al.*, 1992).

Scoping and Issues of Scale in SEA and EIA

A key part of the scoping activity is to define the geographic area(s) for the assessment. This is a key activity as it bounds the areas in which effects are identified and influences the complexity of the assessment. Scoping helps to define the data assembly tasks and the indicators selected for the assessment, as well as aiding the assignment of significance.

As some transportation measures addressing a local problem may only be viable as part of a wider transportation strategy, such as electrification of an entire rail line, so the study area may extend beyond is original geographic focus. The issue is then how the costs and benefits, both financial and environmental arising from such up-grading are allocated within the study as they would be delivered along the length of the rail line, not just within the study area. Essentially, at what point should the assessment cease to consider geographically remote, second order consequences? Also, effective solutions may be available beyond the administrative domain of the organisation investigating the transportation problem.

Difficulties in selecting an appropriate scale for the assessment are magnified where several alternative corridors affecting urban areas are being examined, such as towns along inter-urban routes. The desire may be to focus the transportation model upon the inter-urban movement patterns, and hence devote limited coverage to urban movements.

The size of the study area is also a determinant of the methodologies that can be adopted. However, a large study area does not mean that local area issues should be excluded from the assessment. Rather, the assessment methodologies applied under the area-wide assessment are supported by more detailed methodologies focussing upon those areas considered to be important. These two forms of assessment are:

- *Area-wide assessment:* Examination of the environmental performance of the transportation strategies across the entire Multi-Modal Study area using broad-brush indicators;
- *Local assessment:* Examination of the environmental performance of the individual transportation measures and strategies at a local level. This may comprise a link between two nodes, a local network or single sites.

Various questions can be posed to ensure that the appropriate levels of assessment are undertaken. The elements that should be considered in arriving at such decisions on the scope of the assessment are summarised in Table 14.1.

Table 14.1 Questions to Aid and Clarify the Scope of the Assessment

Task	Comment
Focus on significant impacts	Some transportation measures do not require an EIA and hence are appropriate to consider in SEA. EIA is not good at addressing national or global impacts so assess in the SEA. Identify cumulative effects as project EIAs are often poor at this task.
Some impacts can only be assessed at a later stage	While some impacts cannot be assessed until after decisions are made, their early identification remains important as helps to inform scope of EIA.
Some impacts are the domain of other sectors or SEAs	Regulations may require some impacts to be addressed later in the process or by other organisations.
Some impacts are not significant or are too localised	Thresholds may suggest that impacts are discounted for later assessment, but consider potential for cumulative effects and if in doubt assess it.
Some impacts can only be mitigated at higher decision-making levels.	As well as addressing significant impacts, potential exists to mitigate or avoid many small or indirect impacts.

Source: Adapted from Ferrary, 1992

Approaches to Prediction

SEA demands a high standard of prediction and assessment practice. Firstly there is a need to consider relationships between the plan/programme and existing environmental problems as well as national or international environmental protection objectives. Then forecasts of both the future baseline situation and the environmental performance of the transportation measures are necessary. This raises considerable challenges at a strategic level where the geographic area being examined and the influence of external factors may be great.

Objectives and Indicators

The selection of objectives and indicators can be a process that is open or closed to the public. Often the public has little involvement in the definition or validation of such objectives. This suggests a need to isolate a set of core objectives, indicators and targets that are set nationally. National targets may then need to be apportioned to reflect the scale or characteristics of the study area. A further set of regional or sub-regional indicators may also be required to properly consider the environmental issues associated with the study area and the transportation measures being evaluated.

Clearly, there is a need to be pragmatic in setting the objectives and indicators if for no other reason than the cost and effort needed to assemble the necessary data. As the transportation model will be a key source of data on which to assess the achievement of objectives, so environmental considerations should be taken into account when specifying the transportation model.

There is a risk that indicators for transportation, economic and environmental effects may exhibit some degree of double counting. For example, financial benefits from accident reductions are included within the economic appraisal, while the environmental and social indicators may also consider accidents and community safety issues. Although it is good practice to avoid double counting in different parts of the evaluation process, a pragmatic approach is needed which allows a meaningful exploration of the issues, even where this brings some degree of double counting.

Since the early nineties the OECD has led work on the development of environmental indicators for transport. This work has been taken forward in the US where the Environmental Protection Agency issued a report *Indicators of the Environmental Impacts of Transportation* in 1996. The European Environment Agency has issued reports on indicators under the title *Towards a Transport and Environmental Reporting Mechanism* (EEA, 1999). A common theme running through such indicators is their focus upon state of the environment indicators in order to gauge the effectiveness of broad policy instruments, rather than or the forecasting of change resulting from policy, plan or programme actions.

Indicators need to be fit for purpose. Too often they are developed for one purpose and then transposed for use into other spheres on unsound foundations. Instead, indicators need to be hierarchically organised in a manner that mirrors the information needs associated with the level of decision-making. The need stimulated in part by SEA is for tactical indicators that allow the analyst and decision-maker to distinguish between alternate transportation measures. It should also be recognised that the role of specific indicators may change as a transportation measure passes through the different levels in the transport planning process as the understanding of the environmental conditions or transportation measures and their effects evolve.

Assessment Years

Within project EIA, the time-scales for the assessment year are often no more than a few years ahead of the EIA and the decision. For example, transport EIAs are often based upon an appreciation of the existing environmental conditions, the opening year and design year of a scheme (typically 15 years) with and without the proposal. In the case of SEA of transportation plans and programmes, the assessments may be faced with several transportation measures that may be capable of implementation at different time periods and potentially with a planning horizon some 30 years ahead. This situation makes the task of forecasting the baseline and the 'do something' transportation options fraught with difficult assumptions. For example, the environmental effects are dependent upon the future baseline conditions and such conditions may be altered by transportation measures introduced early in the programme or plan.

With a baseline year and plan horizon year perhaps 20–30 years ahead, the question arises whether intermediate assessment years are needed to address the timing of transportation and environmental effects. It should be noted, however that each additional assessment year multiplies the amount of modelling needed and increases the complexity of the task of identifying the preferred transportation option.

Defining the Future Baseline

The transportation future baseline situation can be defined as the existing situation, plus those changes to the transport system which are currently committed alongside significant land use developments generating new origin-destinations. Potentially, traffic management measures being considered to fulfil air quality management plans and road traffic targets should also be included.

Unfortunately, the situation is made complex given the different ways in which highway and non-highway schemes are programmed. Considerable judgement is needed in defining the public transport elements of the future baseline as bus and rail planning, which largely falls to the private sector, is typified by generally short time horizons with little forward planning.

In addition, it is necessary to define an environmental future baseline. This can become a complex task depending upon the geographic extent over which transportation and environmental effects are anticipated, and the assessment year(s) on which the appraisal is to be based.

Data Needs for SEA

With project EIA, it is generally feasible to undertake field surveys and assemble recent local data to establish baseline conditions. With SEA, however, the study area is likely to be large precluding anything other than reconnaissance surveys. Hence data assembly must be problem-focused, making indicators an indispensable part of SEA.

A clear understanding of the objectives of the exercise should drive data assembly, placing reliance upon a highly discriminatory approach to what environmental information is required at particular levels of detail. Public/ stakeholder consultation should be seen as a key part of data assembly exercises seeking balance in the information sought with the practicalities of cost and time.

The study area needs careful definition as effects may extend beyond the immediate transportation network under consideration. Equally, it should be recognised that an heterogeneous approach is often needed as some areas and topics may require a more detailed understanding than others.

Cumulative Effects

Within transport plans, there is often a strong multi-modal focus in which multiple transportation measures sometimes remote from the target transportation issue may be considered. It is also possible that the transportation measures comprise civil and non-civil engineering solutions implemented within differing time-scales and different areas.

The cumulative consequences of transportation measures upon the different transportation networks are rarely considered. Instead, the focus is upon principal transport corridors to the detriment of effects upon the local road hierarchy and other more diffuse movement patterns taken by public transport, cyclists or pedestrians. This situation arises due to the difficulties associated with data availability, as well as the transport models used. Also the use of thresholds to restrict the analysis may also exclude some cumulative effects.

Significance Criteria

While the practice of defining significance criteria within project level EIAs is often not to the highest standards, the application of such criteria to SEA requires further development. For example questions such as whether there should be a standardised set of significance criteria for SEA that are applied nationally to provide consistency or whether criteria should be developed on an individual plan basis.

Stakeholder Involvement

The SEA Directive proposes that the public should be given an 'early and effective' opportunity to comment on the draft plan or programme and the accompanying Environment Report. It also requires the local authority to inform the public of the adopted plan or programme, reasons for choosing the strategy in light of reasonable alternatives and how environmental considerations and consultation issues have been taken into account. These specific requirements are to be seen alongside the obligations under Article 7 of the Århus Convention on Access to Information, Public Participation in Decision-Making and Access to Justice in Environmental Matters (UNECE, 1998).

It is widely accepted that public involvement can and should be an integral part of the SEA process. Indeed, Therivel and Partidario (2000) argue that participation can bring local knowledge that can improve the design and evaluation of options, as well as help identify mitigation measures to better address people's real concerns. In addition, they argue that public involvement helps to better reflect the plurality of interest and environmental values involved in the development of the strategic actions and to educate all those involved in the process.

There are numerous practical issues associated with stakeholder involvement in SEA that are not found with project EIA. Perhaps one of the key issues is being able to secure stakeholder involvement when most people only become interested when a proposed project directly affects them. Plans and programmes are likely to be too remote to involve people other than members of interest groups.

A further difficulty is the large geographic area that is often encompassed within SEAs which mean that the resources devoted to public involvement are inevitably spread thinly. Outreach methods employed then tend to be biased towards the provision of information rather than the active seeking out of stakeholder involvement from hard to reach communities. The Internet is increasingly being used as an outreach mechanism, but there is a self-selecting exercise taking place leading to the views of only certain interests being represented.

Formal consultation mechanisms remain to be established for SEA, since it is only where the SEA links with an existing statutory process, such as the formulation of development plans that the assessments are currently open to public examination. However in such circumstances public involvement often only commences when the draft plan is made available for formal review (on deposit). Informal consultations are certainly encouraged by government, but practice and attitudes towards public involvement varies from the informed to the hostile.

EIAs are rightly criticised for being too technocratic and it would appear that SEAs for the transportation sector risk the same criticism unless procedures are put in place and effective methods of gaining public involvement are promoted and funded.

The reasons for stakeholder involvement in transport pricing as presented in the chapter by Viegas and Macário, are also valid in the wider domain of transportation planning. Consequently, their conclusion that the acceptability of a pricing strategy results from an interaction between political effectiveness, equity and social fairness and technical feasibility also applies to the formulation of transportation plans, policies and proposals.

Increasingly, decision-making is no longer the domain of a single organisation capable of ignoring dispute. Instead problems must be defined according to the different perspectives of stakeholders and presented in a manner that informs public opinion in order to achieve consensus. This leads to the conclusion that assessment practice needs to be considered in terms of three different levels:

- *Aggregate policy level:* Addressing the economic and social effects concerning economic efficiency, cost of public expenditure, distributional effects and sustainability;
- *Stakeholder level:* Addressing the social acceptability effects as perceived directly and indirectly by various social groups;
- *Practical feasibility level:* Addressing legal, regulatory, technical and managerial issues associated with implementation.

Approaches to Mitigation and Enhancement at a Strategic Level

The assessment of mitigation and enhancement measures within SEA is different from that within a project EIA due to the increased uncertainty and the availability of new mitigation/enhancement measures. Confidence in the successful application of mitigation measures also may influence the robustness of the assessment outcome by altering the assigned assessment scores. Mitigation measures may also provide opportunities for effective responses to cumulative effects. These aspects are explored in the following paragraphs.

Increased Uncertainty

As the SEA is conducted upon conceptual designs that indicate the engineering feasibility of the transportation measures so detailed project and environmental information is not available. In addition, the need for mitigation/enhancement measures is also subject to considerable uncertainty. A further complication concerns whether a particular mitigation measure is capable of being delivered.

Availability of New Mitigation/Enhancement Measures

The mitigation/enhancement measures available at a project level EIA are supplemented within SEA by the availability of institutional measures such as partnerships between organisations, legislative or policy measures. Increased horizontal integration across sectors is expected, as it is probable that parties other than the proponent of the transportation strategy/ measure(s) may be responsible for delivery of some mitigation/ enhancement measures. This opportunity arises from the long lead-time that the SEA affords to the planning processes to address issues of mitigation or enhancement and the fact that a variety of transport infrastructure and service providers are likely to be affected by the plan.

Influence on Assessment Outcomes

The robustness of the assessment may be dependent upon whether mitigation or enhancement measures associated with particular transportation measures can be delivered and achieve a satisfactory level of performance. Where the performance of a particular component of the plan or programme is affected by the provision of mitigation or enhancement measures, then those measures should be developed to a level of detail necessary to confirm that they can be delivered in engineering and financial terms.

Reports should state where mitigation/enhancement and monitoring measures are appropriate and are fundamental to the assessment outcome. Where confidence in the provision of mitigation/enhancement measures would not affect the robustness of the assessment, then opportunities for enhancements and mitigation measures should be recorded within the Environment Report.

Responding to Cumulative Effects

In seeking to mitigate cumulative effects, often the same type of mitigation/ enhancement and monitoring that would be recommended in an EIA is appropriate. However, this may involve measures being applied to actions beyond those that are the centre of the transportation plan or programme. Also, measures may be required to respond to effects remote from the proposed location of individual transportation measures.

While mitigation measures should generally be directed towards the resource being affected or the action causing the cumulative effect, the concept of 'no-net loss' may be an appropriate response and hence some transport measures may include mitigation measures for other projects. Partnerships with local agencies and transport infrastructure or service providers may be the only means of addressing complex cumulative effects as it may be unreasonable to expect a single proponent to bear the burden of mitigating effects attributable to other transport measures in the study area.

The presentation of mitigation measures in both SEA reports and EIAs, like predictions, is an aspect where there is scope for improvement. The following aspects should be considered while specifying the specification of mitigation or enhancement measures:

- clarity in the performance objectives;
- clarity on those responsible for implementation of the measures;
- respect for the subsequent design flexibility that exists;
- awareness of tender specification process to follow;
- recognition of potential contractor practices;
- limit restrictions to be placed on contractors to those absolutely necessary;
- statement of commitments to stakeholders as appropriate;
- clarity of presentation;
- summary of the mitigation/enforcement measures and commitments.

Reporting to Decision-Makers

The purpose of the SEA is to inform decision-makers in a manner that facilitates the evaluation and comparison of alternatives. In contrast, within a project EIA the focus is generally upon whether the environmental effects of the proposed action are likely to be significant and require mitigation.

Within a transportation SEA, many different alternatives may be examined, often proposed for different locations and in different time-scales. One multi-modal study in the UK was initially dealing with approximately 150 separate measures. In reporting to decision-makers, the key information must be distilled and documented in a manner that avoids introducing bias.

In comparing alternatives, an array of different environmental impacts may be measured in monetary or other units as well as in qualitative terms. In such situations, alternatives need to be assessed in terms of their relative performance within topic categories such as air quality, noise etc. In this context, the development of the Appraisal Summary Table represents a step forward.

Tiering between SEA and EIA

The US National Environmental Policy Act 1978 emphasised the importance of advanced planning and preventing decision from being made in cumulatively destructive increments. The Act consequently states that decisions of a higher order must be covered by EIA. The Council on Environmental Quality then gave the following definition to tiering (see Box 1).

Tiering is consequently a process that ensures that the environmental implications of an action are addressed at appropriate levels within the decision-making process and with an appropriate amount of effort to provide robust information for decision-makers. The theory is that SEA and subsequent EIA should be consistent with and reinforce each other, with the former providing the framework for setting the scope of the latter with the prospect of a more focused and efficient EIA.

Box 14.1 Council on Environmental Quality Definition of Tiering

Tiering refers to the coverage of general matters in broader environmental impact statements (such as national program or policy statements) with subsequent narrower statements or environmental analysis (such as regional or basin-wide program statements or ultimately site-specific statements) incorporating by reference the general discussions and concentrating solely on the issues specific to the statement subsequently prepared. Tiering is appropriate when the sequence of statements or analysis is:

From a program, plan or policy environmental impact statement to a program, plan or policy statement or analysis of lesser scope or to a site-specific statement or analysis.

From an environmental impact statement on a specific action at an early state (such as need and site selection) to a supplement (which is preferred) or a subsequent statement or analysis at a later stage (such as environmental mitigation). Tiering in such cases is appropriate when it helps the lead agency to focus upon the issues which are ripe for decision and exclude from consideration issues already decided or not yet ripe.

Agencies are encouraged to tier their environmental impact statements to eliminate repetitive discussions of the same issues and to focus on the actual issues ripe for decision at each level of environmental review. (...) [the] subsequent statement of environmental assessment need only summarise the issues in the broader statement and incorporate discussions from the broader statement by reference and shall concentrate on the issues specific to subsequent action.

Council on Environmental Quality (1992), *Regulations for Implementing the Procedural Provisions of the National Environmental Policy Act*, 1978.

Apart from tiering providing vertical integration, it also has a horizontal component through linkages across sectoral areas, such as those between transportation planning, economic appraisal and land use planning. Hence mechanisms are needed by which information is passed between various sectoral plans and programmes, as well as providing clear boundaries on those topics that perhaps transcend more than one sector.

While integrated assessment specifically seeks to address the horizontal component of tiering at policy, plan, programme and project levels, in practice, the uni-dimensional perspective of transportation planners needs to be challenged in order to bring about integrated assessment. Transportation planners must appreciate that transport is only a means to achieve the wider objectives held by society both at a national and local level. Hence, transportation plans, policies, programmes and projects should be assessed in terms of the extent to which they achieve these wider objectives rather the narrowly focused on addressing congestion. The adage 'it's a transport study' to dismiss other subjects should be seen as outdated in this multi-dimensional integrated transportation planning process.

Barriers to effective vertical tiering between SEA and EIA/project delivery include, institutional frameworks; changing circumstances (and political agendas); and unrepresentative stakeholder involvement. These aspects are explored below.

Institutional Frameworks

Institutional frameworks can provide barriers to tiering since the organisation responsible for the SEA may not have any direct jurisdiction over the projects that are ultimately delivered. Indeed, project EIAs may be undertaken without any requirement to have regard to the conclusions of the SEA. Public examination of the resulting EIA may, however, be expected to be adverse where the EIA is not shown to be in conformity with higher level assessments. This situation could be addressed if the regulations implementing the Directive explicitly state that subsequent project EIAs should be in general conformity with higher level assessments and that the mitigation/enhancement measures identified during the SEA should be implemented.

Unfortunately, project EIAs may be undertaken outside a neat progression from plan, programme to project, since development proposals cannot be delayed until any particular SEA is complete. Hence the land use and transportation context within which the SEA is undertaken will not be static. EIAs may also be underway in parallel to a SEA being undertaken for the same sector or locality. Issues concerning data transfer, commonality of assumptions and precedence then arise. A further issue is that some transportation measures within the SEA would not be classed as projects needing an EIA and hence, mechanisms to ensure conforming environmental performance or environmental monitoring may be lacking.

Changing Circumstances

Tiering also implies a change in attitude at the project level EIA, since it is not appropriate for issues to be re-opened unless it is clear that circumstances have changed from that during the SEA. While this is the theory, given the time lag between SEAs and project EIAs, it is likely that circumstances could have changed. Hence the savings that tiering is envisaged to provide in terms of improved scoping for project EIA may not materialise.

Unrepresentative Stakeholder Involvement

While stakeholder involvement is held up as an important component of SEA, in practice it remains to be seen whether the positions taken to support given strategies during the SEA actually are translated into support at the project level. There are many reasons for suspecting that this is no more than a faint hope. As noted earlier, those involved with the SEA are likely to be an unrepresentative set of stakeholders that are unlikely to include those for whom the projects would directly affect their interests.

A second set of flaws in the view that tiering is beneficial revolves around the assumption that those signing up to a position during the SEA would retain the same position when some of the uncertainties of the SEA give way to the hard realities of project delivery. Again given the prospect of several years between the SEA and EIA, at least for major infrastructure projects, individuals, organisations and communities will change, so previously stated positions may not be retained.

Review, Monitoring and Post-Project Evaluation

While considerable attention is given to issues of procedure, content and methodologies, the mechanisms for the review of Environment reports, monitoring and post-project evaluation have been largely overlooked. One recent publication by Arts (1998) provides some useful insights. In this section, some of the issues associated with review, monitoring and post-project evaluation are explored.

Review

The key issue associated with a review of Environment Reports is essentially one of accountability, so checks and balances need to be in place to ensure that the process is properly applied and maintains public confidence in the integrity of the process.

In the case of project EIAs, it is generally the developer that submits the report to the competent authority in support of an application for consent. Typically, the competent authority has no direct interest in the application and hence should be acting impartially. The situation is, however, different with plans and programmes. It is likely that the competent authority for a policy, plan or programme will also be responsible for undertaking the SEA. In effect the process is that of self-assessment. While there are advantages in this approach in that it promotes long term internalisation of environmental values and should aid informed decision-making, it does raise demands for internal and external mechanisms to monitor performance and verify accountability.

This then raises the question of whether there is a need for an independent organisation capable of scrutinising and reporting on the performance of the plan and programme making authorities in relation to SEA (Sadler and Verheem, 1996). Given that the introduction of SEA will for many institutions represent a change in current practice, the need for clear ground rules to promote good practice that is transparent and open to all is needed.

The defining factors for success with SEA and EIA is the avoidance of unnecessary environmental damage and the successful translation of the mitigation and enhancement measures into legal requirements as planning conditions or project specifications. If the system fails in achieving this then job creation may be its only accolade.

Too often in EIA attention is placed upon the process of assessment and decision-making with too little attention given to issues surrounding the quality of the assessment, the accuracy of forecasts or the effectiveness of mitigation/enhancement measures. Essentially once a decision is taken, the EIA is largely redundant. As a result unless a project is adequately constrained through planning conditions or agreement, the flexibility of project implementation, can result in environmental consequences not considered within the EIA or mitigation actions not being effectively implemented. The lack of effective environmental monitoring to ensure that the desired outcome is achieved leaves much of the EIA process as potentially an expensive charade. Will the same be said of SEA?

Monitoring

Monitoring activities associated with SEA have two purposes. The first is to provide data to ensure that project delivery is in conformity with the approved plan or programme. The second is to identify the monitoring programmes needed during the transportation project planning process e.g. ecological surveys etc. that aid the project assessment process. In this context the key components to be considered are:

- measurable indicators in order that the magnitude and direction of change can be recorded;
- time-scales for the commencement and duration of monitoring;
- appropriate spatial scales over which monitoring is to be undertaken;
- investigation of methods by which causality can be established;
- guidance on appropriate monitoring methodologies;
- consideration of the cost-effectiveness of monitoring;
- institutional arrangements to facilitate monitoring, data assembly and response measures.

Post-Project Evaluation

The performance of environmental assessment is determined by the extent to which it meets established goals or objectives, accepted provisions and principles. This gives a focus on what were the results and what actually happened? Research by Bisset and Tomlinson, (1988) identified that there were many fundamental problems with post project evaluation that make it difficult to assess the accuracy and effectiveness of project EIAs. Among these problems were changes made to the project and changes in the environment that make it difficult to measure and attribute the outcome to the assessment process.

The difficulties with the ex-post evaluation of project are greatly increased with SEA as the chain of cause and effect is often unclear, not made easy by the passage of time between the plan and project delivery let alone any subsequent effects. A study has just been launched between TRL and the University of Manchester to explore such issues.

Recommendations on the Way Forward

While it is perhaps early to be regarding SEA as either providing the salvation for project EIA, or to regard the exercise as being of little real value, it is clear from the above that there are several aspects where further development will be required. These include:

- the need for legislative and procedural linkages between SEA and EIA for transportation projects;
- guidance on forecasting future baseline conditions up to 30 years ahead;
- how to encourage effective public involvement;
- quality control mechanisms.

A key action over the next decade should be to show whether the good words and commitments resulting from SEA and EIA actually deliver an improved environment alongside the improved transportation networks. The UK has long recognised the need for an assessment of plans and policies and with the advent of the SEA Directive and the guidance on SEA for Multi-Modal Studies, the commitment is being translated into action.

Legislative and Procedural Linkages

The UK has embarked upon establishing procedural links between SEA and EIA for transportation projects ahead of considering the legislative requirements. In seeking to link SEA to EIA, it is clear that procedures by which environmental information, commitments are conveyed through the policy, plan, programme, project chain need to be explored. This is particularly important since this view of the planning process rarely matches reality as projects are often generated ahead of or in parallel with plan formulation.

A further dimension is the manner in which SEAs for transport plans link with other sectoral SEAs particularly those associated with land use plans operating in the same geographic area. This may be particularly difficult when transport planning adopts a longer planning horizon than land use plans.

Legislative dimensions may also need to be explored to ensure that project Environmental Statements report that the environmental performance of the project is in conformity with relevant SEAs either for transport or land use sectors. Equally, legislative provisions may be required to ensure that mitigation/enhancement and monitoring measures considered to be essential in the SEA are delivered and that they do not fall between gaps between public administrations and private sector transport operators.

Guidance on Future Baseline

As SEAs for transport plans will operate at a variety of geographic scales, so the environmental attributes of the study area will be examined at different levels of detail. When dealing with urban or sub-regional transportation SEA, then land use and socio-economic change may become important considerations when the future baseline condition is some 20 or 30 years hence. This creates both a technical and political problem in defining the characteristics of the future 'do minimum' situation. When dealing with large geographic areas, then the implications of climate change may also become important, for example in changing agricultural practices or the ecological characteristics of the study area.

In the UK, EIAs for highways make forecasts for 15 years ahead from the opening year of the project. Even under these restricted circumstances, the future baseline at year 15 is often assumed to incorporate no change from the opening year. Consequently, when faced with forecasting 20–30 years hence, new guidance is needed if for no other reason to ensure that base assumptions are internally consistent for both the sector and geographic area for which the SEA is being undertaken.

Guidance on Public Involvement

The public generally has shown little interest in forward planning exercises. Often public interest is stimulated only when their interests are threatened in an obvious manner. Such lack of engagement potentially creates problems for SEA in that it will be for those responsible for the SEA to show that they have taken reasonable measures to promote public involvement. When dealing with large study areas, perhaps a trans-regional multi-modal study then accessing local public opinion can be problematic, but strategies will need to be devised in order to protect against potential legal action from interest groups.

Quality Control

The final aspect for action is that of quality control. EIA practice still suffers from poor quality assessments despite the long record of experience. As this experience is lacking in SEA, the quality of the resulting Environment Reports looks likely to be poor. As SEA may be perceived as slowing the planning process, the implications of a failure to deliver a useful product may result in a perception amongst decision-makers that the entire exercise is flawed and should then be radically reduced in scope.

It is suggested, that effort needs to be expended in rapidly improving awareness of SEA and establishing benchmarks on what constitutes an acceptable Environment Report. This then may deal with the short-term quality control issue associated with its utility to decision-making. The longer-term quality control issue associated with the effectiveness of SEA in forecasting environmental effects of transport strategies also needs to be considered in terms of monitoring and post-development audits in order that the utility of the entire SEA exercise can be subject to quality control checks.

Note

1 The DETR has subsequently been reorganised with land-use planning being the responsibility of the Office of the Deputy Prime Minister while transport rests with a new Department for Transport.

References

Arts, J. (1998), *EIA Follow-up: On the Role of Ex-Post Evaluation in Environmental Impact Assessment*, GeoPress, Groningen, The Netherlands.

Au (1999), *Examples of Strategic Environmental Assessment (SEA) in Hong Kong*, Environment Protection Department, Hong Kong.

Bisset, R. and Tomlinson, P. (1988), 'Monitoring and Auditing of Impacts', in Wathern, P. (ed.), *Environmental Impact Assessment*, Unwin Hyman, London, pp.117–128.

DETR (1998a), *New Approach to Appraisal*, DETR, London.

DETR (1998b), *A New Deal for Transport*, DETR, London.

DETR (1998c), *Understanding the New Approach to Appraisal*, DETR, London.

DETR (2000), *Guidance on the Methodology for Multi-Modal Studies*, DETR, London.

EEA (1999), *Towards a Transport and Environmental Reporting Mechanism (TERM) for the EU* European Environment Agency, Copenhagen.

Ferrary, C. (1992), *Environmental Assessment for Policies, Plans and Programmes*, European Transport, Highways and Planning, PTRC XXth Summer Annual Meeting, Proceedings of Seminar B, 14–18 September 1992, Institute of Science and Technology, Manchester.

Partidario, M.R. and Therivel, R. (2000), 'Perspectives on Strategic Environmental Assessment', in Partidario, M.R. and Clark, R. (eds), *Perspective on SEA*, CRC Press, Florida.

Sadler, B. and Verheem, R. (1996), *Strategic Environmental Assessment: Status, Challenges, and Future Directions*, Ministry of Housing, Spatial Planning and the Environment, The Netherlands.

Therivel, R., *et al* (1992), *Strategic Environmental Assessment*, Earthscan Publications Ltd, London.

Tomlinson, P. (1999), 'Strategic Environmental Assessment Guidance in the United Kingdom', paper presented at the OECD-ECMT Conference on Strategic Environmental Assessment for Transport, 14–15 October, Warsaw, Poland.

Webb, J.W. and Sigal, L.L. (1992), 'Strategic Environmental Impact Assessment in the United States', *Project Appraisal*, Vol. 7, No. 3, pp.137–141.

Wood, C. (1992), 'Strategic Environmental Impact Assessment in Australia and New Zealand', *Project Appraisal*, Vol. 7, No. 3, pp.43–149.

World Bank (1993), *Sectoral Environmental Assessment*, Environmental Assessment Source Book, Update, No. 4, Washington DC.

World Bank (1996), *Regional Environmental Assessment*, Environmental Assessment Source Book, Update, No 15, Washington DC.

UNECE (1998), *Convention on Access to Information, Public Participation in Decision-Making and Access to Justice in Environmental Matters*, United Nations Economic Commission for Europe Committee on Environmental Policy, Fourth Ministerial Conference, Environment for Europe, 23–25 June, Århus, Denmark.

Index

Notes: bold page numbers indicate tables and diagrams; numbers in brackets preceded by *n* are note numbers.

acceptability 151, 154, 159–68, 216–19
accessibility 91–2, 148
  analysis 227–42
    applications of 239–41, **240**
    deterrence functions 230, 231–2
    importance underestimated 227–8
    methodologies 232–6, **235**
      opportunity measures 233–4, **235**, 239, 241
      simple measures 233, **235**, 239, 241
      value measures 234, **235**, 241
    models 229–30, **229**, **230**
    need for 228–30
    opportunity terms 230–31
    uses 236–7
    zones 230, 232, 239, 240
  and infrastructure investment 111–12, **111**, 113, 116–17, 121
accountability 197
aesthetic values 139, 140
air pollution 13, 48–9, 65, 83, 142, 173
  see also CO_2 emissions
air transport 19, 43
  agency
    in Norway 22, 27, 28–9, 30, 40
    in Sweden 31, 32, 36, 37
  cartels in 141
Amsterdam 92, 96
  orbital motorway 115
  PBKAL project 125, 210(*n*1)
analysis
  cost-benefit *see* cost-benefit analysis
  level/scope of xiv-xv
  qualitative 88, 183, 192
  quantitive 184
appraisal *see* evaluation
Appraisal Summary Table (AST) 10–11, **16**
Aquitaine TGV 200
argumentative model 222, 224
Aschauer, D.A. 45–6, 47, 90–91
ASEED Europe 59
ASTRA model 149(*n*2)
audits xvi
Austria 134, 202

Barcelona-Montpellier TGV 200–201, 202, 210(*n*1)
Belgium 202
  PBKAL in 131, **131**, **132**, 133
Benelux countries 54, 210(*n*1)
Bergen 73, 76, 78–80
Betuwe railway line 209–10(*n*1)
'bicycle theory' 61
Blue Banana 54–6, **55**, **56**
Böge, S. 53
bottlenecks *see* congestion
Brenner project 134, 209(*n*1)
Brent, R.J. 9–10
British Rail 9–10
Brussels 125, 210(*n*1)
Budapest 65
Buffalo Light Rapid Rail Transit (LRRT) 115

carbon monoxide *see* CO
cars
  emissions 33, 34, 39, 110, 139
  ownership 158
  and public transport 143–5, 213
cartels 141
CBA *see* cost-benefit analysis
central business districts (CBD) 110
Channel Tunnel 59, 109
charges, user *see* road user charges; toll roads
Christophersen list 199
Citizen Network, The (EC Green Paper) 141, 213–14
civil aviation *see* air transport

CO_2 emissions 110
levy on fuel for 33, 34, 39
targets for reducing 139, 143
coastal transport agency (Norway) 29, 30, 40
Cobb-Douglas specification 97
Cohesion Fund 50, 56–7, **57**, 66
Cologne 125, 210(*n*1)
Common Agricutural Policy (CAP) 62
Common Transport Policy 47, 213–15
communication in decision process 14
congestion 48–9, 51, 53, 75, 90
charging 237
variable in evaluations 10, 103, 223
corridors *see* Pan-European Corridors
cost-benefit analysis xvi, 3, 8–9, 21, 184, 192
advances in 94, 145–9, **146**
conceptual foundations of 151–69
model 165–8
in strategic planning 21, 24, 40
surplus measured by 153–6, 157, 159–61
traditional 17, 45–6
problems with 18, 163–4
in transport investments 51, 118, 120
variations in 13, 14
in welfare-based evaluation 139, 140, 142
creative learning capacity 115–16
cross-sectoral planning process 26–7, 29, 32
cyclists 183, 185, 229, 234, 239, **240**

Denmark 121–2(*n*2), 202
strategic planning in 17, 19, 21, 25–6, **25**
distributive impacts 13
Docklands 8, 12
Doherty, A. 59–60
double counting 13, 118, 121, 158–9
Dupuit, James 168(*n*10)

East London River Crossing project 8, 12
Eastern Europe 45, 56, 63–7
econometric model 8–9, 11
economic analysis 3–4, 35
see also cost-benefit analysis
economic geography models 87, 94
economic growth 43, 50, 51, 58, 107–23, 158
economic potential concept 92
economies of scale 54
Edinburgh 240–41, **241**
EIB *see* European Investment Bank
employment *see* jobs
environment
Agency, European 252
evaluating impacts on 4, 19, 53, 65, 179
in decision-making process 197, 198, 203–4
estimating switch value 142–5, **144**
infrastructure investments 108, 110
in specific projects 8, 10–11, **16**
see also strategic environmental assessment
MOCPF in analysis of 147
policies for, discount rates 147
stakeholder interest xvii
strategic assessment of xvii
in strategic planning 18, 22, 24, 28, 31, 34
sustainability of xiv
targets for 173, **193**
pressure on 142
environmental impact assessment (EIA) 186
Directive 244, 245
and SEA *see under* strategic environmental assessment
Environmental Policy Act, US (1978) 257
Environmental Protection Agency 252
equity 218–19
equivalent gain (EG) 169(*n*16)
Essen Council 58
European Commission 47, 51, 60, 62–3, 65, 133, 135
on public transport 141, 213–14
European courts 8
European Environment Agency 252
European Investment Bank (EIB) 47–8, 58, 66, 133
European Regional Development Fund (ERDF) 47, 57
European Round Table of Industrialists (ERT) 59–60, 61, 199
European Union (EU)
Common Transport Policy 213–15
cross-border projects in 125–38
funding 47–8, 61
Cohesion Fund 50, 56, **57**, 66
modal bias 56–8, **57**
see also Trans-European Networks
Gothenburg Summitt 49
interventions in Central Europe 43–71
legal obligation 46–7
regional infrastructure **52**
regional policy 47
see also regional development
evaluation
accessibility analysis *see* analysis *under* accessability and decision-making, relationship between 7–12

effectiveness of xiv-xv
environmental *see* evaluating *under* environment
European versus national-level 125–38
problems with 125–8, 134–6
Lost & Found 126–8, **127**, 131, 135, 136(*n*1)
solutions to 128–31, **130**, 134
ex ante 17, 19
ex-post *see ex post* evaluation
of infrastructure investments 86–96, 108, 117–20
methodologies 17
multi-stakeholder 14
in the policy process xvi-xvii, 3–16
in policy process *see* policy process
of pricing policy 219–24
frameworks for 221–4
process 5–7
role of xv-xvi, 1–84
social impact assessment 187, **187**, 190
strategic planning 17–42
technical aspects xvi, 85–169
role of 3
welfare basis of 139–50
switch values in 142–5
ex ante evaluation 17, 19
ex post evaluation xv, 9–10, 121
in Scandinavia 17, 19, 23, 24–5, 26–37, 190–1

Femer Belt project (Denmark) 21
Finland
infrastructure described 174–6, **175**
road administration, strategic management of 173–95, **177**, **180**
documentation and audit 188–9
evaluation methods 183–9, **187**
impact assessment 187, **187**, 190, **195**
objectives 189–90, **194**
targets 173, **193**
SEA in 179
Finnra 173–92
fiscal pressure 145
FISCUS research project 217
France 54, 135
decision-making process in 203, 205–206
PBKAL project in 131
PREDIT programme 199
rail projects in 200–201, 202
freight transport 33, 53
and fuel taxation 142
and infrastructure investment 110, 112
Fréjus tunnel 200
fuel
consumption 83, **83**, 143–4, **144**
taxation 142

Geographical Information Systems 240
Germany 53, 54
Länder 46, 49
PBKAL project in 131, **131**, **132**
Gothenburg European Summitt 49
government 5, 14, 18, 39, 157, 227
'capture' of 141–2
decision-making process of 197–209
local 140, 142, 201–203, 204, 217
regional 31–2, 38, 39
transparency 11, 29, 39, 156
see also individual countries
government bodies xiii
Great Belt project (Denmark) 21
Greece 57
inter-island network project 210(*n*1)
green parties 197
Greenwich, London Borough of 8
Groningen 87, 96–7, **97**, 100
Growth, Competitiveness & Employment (EC White Paper) 47

Hamburg 92
Heathrow Airport 115, 164
Helsinki (region) 175
Helsinki Corridors 56, 62, 63–4, 66
high speed trains 56–7, 59, 87, 125–38, 200–1
Hoedeman, O. 59–60
Hungary, rail system 63–4, **63**, **64**

iceberg type costs 98, 100–101, 103
impact assessment, social **187**, 190, **195**
Industrial Project Analysis, Manual of 142
industry 31
information provision xiv, 204–206
information technology (IT) 175
infrastructure assessment 43–71
infrastructure investments 46–56
decision-making process 197–211
economic impacts of 87–105
causality 108–12
conceptual model **93**
estimating, methods for 89–96
models 94–6
macroeconomic aspects 44, 87, 92–4

microeconomic analysis 90–92, 120
national effects 100–101, **100**
regional effects 91–100
economic growth 107–23, **111**
conventional view 108–10, **109**
developed view 116–20
misconceptions 113–14
policy design and evaluation 115–20
spacial patterns 101–103, **102**
types 88–9, **88**
political implications 45–6, 58–62
transport costs 44–5
Infrastructure Use, Fair Payment for (EC White paper) 214
institutional basis for evaluation xiii
Inter-Regional Trade Theory 45
intermodal transport 29, 112
International Passenger Survey 126
Italy 51, 54, 60
environmental targets in, evaluating 143

jobs
creation 48, 147
distribution 99–100, 233
and investments, impacts of 110
journey time *see* travel time
just-in-time delivery 43, 112

Köln 125, 210(*n*1)

land use 53, 148, 198, 228, 238
land-use/transportation interaction (LU/TI) model 87
compared with SCGE models 94–6, 103
landscape/aesthetic values 139, 140
LGV Nord 125–6
Lindblom, C.E. 5–6
lobby groups 4, 59–62, 199
local authorities *see* local *under* government
London 12, 54, 115, 164
local government 8
PBKAL project 125
London Airport Inquiry 3
LU/TI model *see* land-use/transportation interaction model
Lyon-Turin TGV 200–201, 202, 204, 210(*n*1)

Maastricht Treaty 60, 145, 199
Mackie, P.J. 11
marginal cost principles 32
Marginal Opportunity Cost of Public Funds (MOCPF) 145–8, **146**
Martin, Philippe 45, 54
methodologies 19–26, **25**, 38–40
development of 40
Mezzogiorno region 51
Ministry of Transport 4
MOCPF *see* Marginal Opportunity Cost of Public Funds
mode-by-mode approach 17–18
models
econometric 8–9, 11
'muddling through' 5–6, **6**
rational/analytical 5–7, **6**, 11
spacial economic 87–8, 94–9, 101, 103
transport demand 32
monetised/non-monetised information 11, 13
monopolies 141, 148
Mont Blanc tunnel 200
'muddling through' model 5–6, **6**
multi-criteria analysis 4, 21, 24, 140, 142
multiplier analysis 119
multiplier effects 108–109

Nellthorp, J. 11
Net Present Value (NPV)/Cost ratios 9
Netherlands 90
government 68(*n*11)
PBKAL project in 131, **131**, **132**
rail infrastructure 89, 92
evaluation of proposals 96–9
Newbury by-pass 4
Nilsson, J.E. 8–9
noise pollution 11, 13
Non-government organisations (NGOs) 29
Norway
government 22–3, 26, 28, 75
legislation 73
local 39, 73, 83
Ministry of Transport 73
National Roads Administration 80
road tolling 73–84
strategic planning in 17, 19–20, 21–3, 37–40
ex post evaluation in 26–31
National Transport Plan 2002–2011 28–31, 38
Road Traffic Plan 1998–2007 26–8, 38
methodologies 25–6, **25**
National Transport Plan 2002–2011 22–3
NPV/Cost ratios 9

objective-oriented planning 18, 20, 21–2, 27

OEEI (Netherlands) 7, 13
opinion polls 79, 90, 126
Øresund Link 21, 59, 109, 121–2(*n*2), 129, 209(*n*1)
 decision-making process 200, 204
Oslo xvi, 73, 76, 78–80, 83
 region 73–4, 80

PA (project analysis) 8–9
Pan-European Corridors 56, 62–4, **63**, **64**, 65, 185
Pareto welfare gain 148
Parsons, Wayne 3
PATS research 216, 218
Pavement Management System (PMS) 184
payment mechanisms 159–63
 pay-as-you-go principle 161–3, 164–5
PBKAL project 125–38, 210(*n*1)
pedestrians 183, 185, 228, 229, 234, 239, **240**
 walking threshold 233
Phare Program 62
planning, cross-sectoral 26–7, 29, 32
Planning Inquiries xvi
policy process
 accessability issues in 237–9
 evaluation in 171–264
  stakeholder involvement 213–25
policy-oriented planning process 26–7
political criteria xv-xvi, 3–4, **4**, 28, 36, 121, 134–5
 growth of TENs 59–62
 of infrastructure projects 197–211
 strategic planning 37–8
 in welfare-based evaluation 140, 147
'polluter pays' principle 143
pollution 48–9, 65, 83, 142, 173
 in analysis 11, 13
 see also CO_2 emissions
PREDIT programme 199
pricing
 assessment of xiv, 19, 135, 142
 policies 213
 stakeholders' involvement in 213–25, 255
 evaluation frameworks 219–22
process-oriented approach 24
project analysis (PA) 8–9
project objectives xiv
project-oriented planning 21, 37
public capital hypothesis (Aschauer) 45–6, 47
public choice theory (PCT) xvi, 139, 140–45
Public Inquiries 8, 12
public opinion 3, **4**
 acceptability 216–19
  equity principle 218–19
 consultation process 4, 206–209
public policy analysis 197
public-private partnerships (PPP) 5, 199, 200–1, 209
public transport
 alternative to cars 27, 143–5
 funding 73
  subsidies 33
 urban
  accessability 227–42
  pricing 213–15
 welfare-based evaluation of 141, 143, 158–9

qualitative analysis 88, 183, 192
quantitive analysis 184
 see also cost-benefit analysis
quasi production function approach 91
questionnaires *see* opinion polls

RAEM model 97–9, 101, 103
rail agencies 19
 Denmark 21
 Norway 22, 27, 28–30
 Sweden 31–2, 33, 37, 39
rail investments 45
 British Rail 9–10
 compete with other modes 43
 economic impact study 96–9
 employment effects 48
 and TENs 56–8
rail links, high speed *see* high speed trains
Ramsey-pricing scenario 145
Randstat, Netherlands 96, 100
rational/analytical model xv, 5–7, **6**, 11, 224
recession 12
regional development 43
 see also individual countries, and *see* local *under* government
 in strategic planning 18, 22, 24, 31, 33–4, 37, 39
 targets **193**
 territorialism 201–203
 and transport investments 45–56, 91–100, 107–23
Regional Trade Theory 45
regulatory measures, assessment of xiv, 19
residents' interests xvii
Rhône-Alpes region 204

Richardson, T. 60
River Thames Bridge 8
road agencies 19
  Denmark 21
  Norway 22, 27, 28–30
  Sweden 31–2, 37, 39
road safety, in strategic planning 18, 24, 28, 31, 33–4
road schemes xiv, 32, 45, 128
  compete with other modes 43
  economic benefits 50
    employment effects 48
  and TENs 56–8
road supply 227–8, 238
road user charging 4, 36, 142
  congestion 237
  public opinion and 3, 216
  see also toll roads
'road-pricing' 80–83
Rotterdam 96
Round Table of Industrialists *see* European Round Table
route choice 74
rural development in strategic planning 28

SACTRA (Standing Advisory Committee on Trunk Road Assessment) 7, 13, 53–5, 228
safety 48, 82, 142
  road, in strategic planning 18, 24, 28, 31, 33–4
  in strategic planning 22, **193**
Scandinavia 200
  differences between countries in 19, 20–21
  government in 18
  regional infrastructure 51, **52**
  strategic planning in 17–42
    methodologies for 19–26, **25**
    road planning 17, 173–95
Scanlink project 59
Schiphol Airport 87, 90, 96–7, **97**
scoping 248, 250–51, **251**
Scotland, accessability analysis in 238–9, 240–41, **241**
SEA *see* strategic environmental assessment
sea transport 43
  agency for 19
    Norway 22, 27, 28
    Sweden 31, 32, 33, 34, 37
Self, P. 3, 5
sensitivity analysis 142
service sector 43
Simon, H.A. 5–7
Simon's rational/analytical model xv
Single Market Review 149(*n*4)
social exclusion/inclusion 13, 158, 238
  see also accessability
social impact assessment 187, **187**, 190
socio-economic (most efficient) alternative 24, 31, 32
spatial computable general equilibrium (SCGE) models 87–8
  compared with LU/TI models 94–6, 103
  evaluation using 96–9, 101
Spain 46, 58, 200–201, 202
speed limitation 33, 34, 39, 142
spillover benefits 45–6, 47, 54
stakeholders xvii, 117
  evaluation of 14
  involvement of 22, 29, 186–7, 189, 213–25, 248
    in SEA 254–5, 259–60
Standing Committee on Trunk Road Assessment (SACRA) 228
static equilibrium models 94
strategic environmental assessment (SEA) xvii, 62–6, 177–9, 186, 243–63
  data needs 253
  defined 243
  Directive 244–5
  and EIA 246, 248–9, 257
    differences between 246, 249, 252, 253, 254–5
    scoping 248, 250–51, **251**
    tiering between 257–60
      defined **258**
  global impact of 243
  methodology 246–9, **247**
  mitigation and enhancement 255–7
  monitoring 261
  post-project evaluation 261
  prediction, approaches to 251–4
  review 260–61
  stakeholder involvement 254–5, 259–60
supply and demand 227–8, 229, **229**, 238, 245
sustainability, environmental xiv, 204
sustainable development 18–19, 58, 181–2
Sweden
  government in 24, 31, 35–6
  projects in 8–9, 121–2(*n*2)
  strategic planning in 17, 19–20, 23–5, 37–40
    ex post evaluation in 31–7

methodologies 25–6, **25**
Swedish Institute for Transport and Communications Analysis (SIKA) 24, 31, 32, 35–6
switch values 142–5
Switzerland 54
system analysis 20
system dynamics 94

taxation 145, 147
technical evaluation xiii-xiv, 3, **4**
telecommunications 44, 110, 175, 240
telematics 185
TENASSESS research 199, 209–10(*n*1)
TENs *see* Trans-European Networks
territorialism 201–203
Thames, River 8, 12
toll roads xvi, 51, 73–84, 159, 202
design 76–8, **77**
environmental benefits 82–83
price structures 81–2, **82**
public attitudes 78–80
revenue raising 75–6
as 'road-pricing' 80–83
traffic
flow, in strategic planning 22, 28
freight 33
passenger 33
safety *see* safety
trains, high speed 56–7, 59
Trans-European Networks (TENs) xvi, 43–43–44, 47–8, 56–62, 116, 209
projects **61**, 65, 125–38, 200–201
TENASSESS analysis 199
TRANS-TALK conferences xiii-xiv, 3
transparency 11, 29, 39, 156, 217
transport
agencies 35
conflicts between 29–30, 38
corridors 33
costs 44–5, 98, 103, **193**
iceberg type 98, 100–101, 103
demand 227–8, 238, 245
modelling 229, **229**, **230**
infrastructure *see* infrastructure investments
investments *see* infrastructure investments
policies 3–16
reliability xiv
Transport Infrastructure Needs Assessment (TINA) 62–6
travel time
savings xiv, 53
as variable in evaluations 10, 11
Trondheim 73, 76, 78–80
Turro, M. 61–2
Twente Central Canal Connection 210(*n*1)
typology of planning and evaluation 20–21

UK
employment patterns 110
Government 13
Department of Transport 8, 10, 150(*n*14)
PKBAL project in 131, **131**, **132**
road user charging in 4
Roads Review 1997–8 10–11, 12, 245–6
Scotland 238–9, 240–1, **241**
SEA in 244, 262–3
supply/demand study in 228
UNITE research project 217
United States 44, 45, 91, 110, 114, 252
unmotorised travel *see* cycling; walking
urban development 43, 173
Urban Public Transport *see* urban *under* public transport
urban transport planning 18
utilitarianism 168(*n*10)

VagverketVagverket Investment Programme 8–9
value added 118–19, **119**
value for money 12
Vienna 62

walking *see* pedestrians
Warsaw 65
'weight' 9–10
welfare 151
evaluation of xvi, 139–50
conceptual foundations of 151–69
and infrastructure investment 111, **111**
revealed welfare assumption 151–3
welfare-theoretic approach xvi
Williams, A. 3, 5
willingness-to-pay/accept (WTP/WTA) 151, 154, 159–68

T - #0098 - 160425 - C0 - 216/150/16 - PB - 9781138708242 - Gloss Lamination